Design Pedagogy

Design Pedagogy

Developments in Art and Design Education

Edited by
MIKE TOVEY

Routledge
Taylor & Francis Group

LONDON AND NEW YORK

First published in paperback 2024

First published 2015 by Gower Publishing

Published 2016 by Routledge
4 Park Square, Milton Park, Abingdon, Oxon OX14 4RN

and by Routledge
605 Third Avenue, New York, NY 10158

Routledge is an imprint of the Taylor & Francis Group, an informa business

Publisher's Note
The publisher has gone to great lengths to ensure the quality of this reprint but points out that some imperfections in the original copies may be apparent.

Gower Applied Business Research
Our programme provides leaders, practitioners, scholars and researchers with thought provoking, cutting edge books that combine conceptual insights, interdisciplinary rigour and practical relevance in key areas of business and management.

British Library Cataloguing in Publication Data
A catalogue record for this book is available from the British Library

Library of Congress Cataloging-in-Publication Data
Design pedagogy : developments in art and design education / [compiled] by Mike Tovey.
    pages cm
  Includes bibliographical references and index.
  ISBN 978-1-4724-1598-1 (hardback) -- ISBN 978-1-4724-1599-8 (ebook) -- ISBN 978-1-4724-1600-1 (epub) 1. Industrial design--Study and teaching. 2. Design--Study and teaching. I. Tovey, Mike.
  TS171.44.D474 2015
  745.2071--dc23

                                                                    2014031286

ISBN: 978-1-4724-1598-1 (hbk)
ISBN: 978-1-03-283703-1 (pbk)
ISBN: 978-1-315-57669-5 (ebk)

DOI: 10.4324/9781315576695

Contents

List of Figures

List of Tables

List of Contributors

Michael Tovey

Michael Tovey is Professor of Industrial Design, and Reader in Design Pedagogy at Coventry University. He joined the institution following a period of practice in industry and was responsible for the establishment and development of transport design at Coventry. This has now achieved international prominence and centre of excellence status. He was Dean of the Coventry School of Art and Design for 18 years. During this time it doubled in size, incorporated performing arts, achieved strong research rankings and a very positive identity and profile.

He pioneered design research, contributing to publications and holding a number of research council grants. He has served on research council committees, supervised and refereed grants and publications.

Much of his research work has been concerned with the design process and how designers work. There has been a particular focus on the use of computer support for the creative aspects of design. The context for this work has been concept design in the automotive industry and the development of novel techniques to support the design activity. He also pioneered the portfolio Ph.D. and this work is a useful summary of much of his research activity.

From 2007 to 2010 he was Director for Design and responsible for leading and co-ordinating design education and design research across the university. In addition to his cross-university role he was Director of CEPAD (the Centre of Excellence in Product and Automotive Design). It has strong links with the Industrial Design Department, as well as connecting with other parts of both the Coventry School or Art and Design, and the rest of the university.

He is a member of the Council of the Design Research Society and leads its Special Interest Group in Design Pedagogy.

Tim Ball

Tim Ball is Senior Lecturer, Industrial Design, Coventry School of Art and Design. He is a designer and educator in product and industrial design with a broad range of subject expertise and interests. From people to emergent materials and processes, architecture to pocket-sized items, there are many influences on his thinking; exploring the exciting, collaborative spaces *between* disciplines that yield new thinking. Design practice and collaboration has forged an extensive network that informs, supports and inspires his teaching. He enjoys the interplay between understanding experiences, the flux of design teaching, the quest for new knowledge, meanings and learning *with* his students.

Erik Bohemia

Erik Bohemia is a Senior Lecturer at the Loughborough Design School, Loughborough University, United Kingdom. Dr Bohemia's ongoing research interest is examining product development processes in geographically distributed and cross-cultural product development teams. The results from his research in this area have been used to guide the development of curriculum in design so that future graduates may more effectively fulfil industry requirements. Dr Bohemia's research has been published in international journals and conferences. For more information visit http://theglobalstudio.eu/

Karen Bull

Karen Bull is Associate Head of Student Experience within the Department of Industrial Design at Coventry University. She is responsible for pedagogic development and teaching of design research methods, critical and creative thinking at all levels of higher education study. Her Ph.D., Advanced Personal Telecommunications Products and Industrial Design, is where she developed significant interest in industrial design methodology and design research pedagogy. Karen was Deputy Director for the Coventry University CETL Centre of Excellence in Product and Automotive Design (CEPAD) and focused on developing an understanding of higher education transformative learning experiences within the field of industrial design.

Ian Campbell

After graduating from Brunel University in 1985, Ian Campbell worked as a design engineer, first in Ford Motor Company, and later in the Rover Group. In 1989, he was appointed as a Senior Teaching Fellow for CAD/CAM at the University of Warwick, where he undertook a part-time M.Sc. degree by research. In 1993, he obtained a lectureship at the University of Nottingham and gained his Ph.D., again through part-time study, in 1998. His current position, since October 2000, is Reader in Computer-aided Product Design at Loughborough University in the Design School. Dr Campbell is editor of the *Rapid Prototyping* journal.

Linda Drew

Professor Linda Drew BA (Hons) MA Ph.D. FRSA FDRS has been Deputy Director at the Glasgow School of Art since 2011. As the most senior academic at the school, she provides leadership in learning, teaching and research. She is currently Chair of Council for Higher Education in Art and Design. Her research focuses on learning and teaching in an art and design practice with both a phenomenographic approach and a social constructivist outlook. Linda's Ph.D. is in educational research from Lancaster University.She is founding editor of the highly regarded peer-reviewed research journal *Art, Design and Communication in Higher Education*, published by Intellect books for over 10 years.

Chris Evans

Chris Evans is Programme Director, M.Sc. Product Design at Aston University, Birmingham. He has created and taught a wide variety of innovative design courses in different UK Universities since 1989. These have been underpinned by 20 years of professional design and management experience with leading manufacturing organisations – designing toys, consumer electronics, transport systems, vehicles, cookware, kitchen equipment, tableware, glass and ceramics. Chris's teaching and practice reveal his core design philosophy: that successful design can only come about through a melding of creativity and innovation with an appropriate engineering and technological knowledge plus an effective understanding of markets and user requirements.

Mark Evans

Mark Evans is a Reader in Industrial Design and leader of Loughborough's Design Practice Research Group. Prior to joining the University he was a corporate/consultant designer with clients including British Airways, Bosch and Honda. His research focuses on supporting design practice through the development of tools/resources and impact of digital technologies on creative practice. Overseas appointments include International Scholar, Massachusetts Institute of Technology and visiting professor, Rhode Island School of Design. Research funding has been received from organisations including the Department of Trade and Industry, Industrial Designers Society of America, Research Councils UK, Hewlett Packard and Royal Academy of Engineering.

Steve Garner

Steve Garner has led some of the United Kingdom's most innovative and popular design programmes, most recently as professor of design at The Open University and formerly as programme leader for Industrial Design and Technology at Loughborough University. Now working as an education consultant he has examined design at secondary, undergraduate, masters and Ph.D. level. Originally a designer in the furniture industry he has contributed to the development of e-learning in design, the use of sketch representations in design, usability in product design and computer-supported collaborative designing.

Aysar Ghassan

Aysar Ghassan's role as a teacher of automotive design at Coventry University is underpinned by his research and his experience as a multifaceted practitioner. He is passionate about integrating the delivery of practical skills and theoretical discourse in design education. Aysar's research into peer-tutoring and professional identity formation reflects the need to broaden the curriculum to prepare students for the global nature of contemporary design practice and the wider knowledge economy. His research interests also include contributions to the development of a contextual understanding of both user-experience research and design for sustainability. Furthermore, Aysar writes design-focused journalistic articles aimed at disseminating research to a broader audience.

Eddie Norman

Eddie Norman is Emeritus Professor of Design Education at Loughborough Design School (LDS). His research concerns the relationship of technologies and designing in general and higher education, and associated pedagogical issues. He led the Design Education Research Group and supervised seven successful Ph.D. students. He contributed to teaching on LDS' undergraduate and masters programmes. He has edited IDATER (1998–2002), D&T Association Conferences (2002–009); and from 2005, *Design and Technology Education: An International Journal*. Prior to joining LDS he had careers in secondary education and as a welding research engineer. Since retiring he has co-founded Loughborough Design Press with Ken Baynes.

Jane Osmond

Jane Osmond is a Research Fellow for the Centre of Excellence for Product and Automotive Design (CEPAD) at Coventry University. Her current research includes pedagogy in relation Art and Design, using the threshold concept theory; the development of an EU transport passenger measurement tool as part of an FP7 project, and researching gender and public spaces, with a focus on public transport. She was awarded her Ph.D. through published work in 2014.

Eujin Pei

Dr Eujin Pei is a Senior Lecturer in Product and Furniture Design at De Montfort University in the United Kingdom. His primary research focuses on design representations and additive manufacture. He worked as a Research Fellow at leading institutions including Loughborough University, Brunel University and the University of Southampton. Prior to joining DMU, he was a product design consultant undertaking work for Motorola Inc., LM Ericsson, and Rentokil Initial. Dr Pei is a Fellow of the Royal Society for the Arts, Manufactures and Commerce (RSA), and an Associate Editor for the *Journal of Assembly Automation*.

Seymour Roworth-Stokes

Seymour Roworth-Stokes is Executive Dean and Professor of Design at Coventry School of Art and Design. He is Chair of the Design Research Society and a Strategic Reviewer for the Arts and Humanities Research Council. An industrial designer, his research explores our understanding of experiential knowledge generated through creative practice and how it can lead to improvements in organisational performance. In 2012, he was presented with the PODIUM Gold award on behalf of LOCOG for the best Higher Education Cultural and Creative initiative of the London games after leading a £2 million consortium project to generate artwork for the Olympic Park.

Alison Shreeve

Professor Alison Shreeve is Head of School, Design, Craft and Visual Arts at Buckinghamshire New University and previously the Director of Creative Learning in Practice Centre for Excellence in Teaching and Learning (CLIP CETL) at the University of the Arts London.

She has a Masters in Art Education and a Ph.D. in Educational Research. Research interests include the student and tutor experience in creative arts higher education. She has published articles in international journals and contributed to several books. She is associate editor of the journal *Art, Design and Communication in Higher Education* and a National Teaching Fellow.

Reviews for
Design Pedagogy

Design research has led to a deep understanding of the nature of design that will be important in guiding design practice in the 21st century. In parallel a growing body of research into design education is having a similar impact on design pedagogy. This book presents excellent examples of such research with wide application across design education.

Chris McMahon, University of Bristol, UK and
immediate Past-President, the Design Society

This is an excellent and timely contribution to the development of our understanding of design teaching. As leader of the Design Research Society's Design Pedagogy special interest group, Mike Tovey has brought together several of its leading members to crystallize thinking in the field. Research in design pedagogy is flourishing and this book provides a significant contribution to the debate.

Tracy Bhamra, Dean of Loughborough Design School,
Loughborough University, UK

Introduction

In our universities and colleges there is a long tradition of teaching design through design practice. For most students their end goal is to achieve a level of capability to function as designers in the professional world. In order to reach this standard students need to achieve a level of professional 'polish' and presentation to match that of the practising designer. However, it can also be argued that the key to their doing this lies in their abilities to think in a solution-focused way employing visuospatial intellectual abilities. Most particularly if they cannot think creatively they will not achieve the required standard. The ability to engage in the creative synthesis of ideas through design thinking is essential if they wish to gain entry to the community of professional practice.

Today it is vital that their education helps them construct a 'passport' to enter this professional group. For many design students the physical manifestation of their passport to design practice is their portfolio of design work. It is in this assemblage of work that they demonstrate that they can tackle design problems to a standard that is recognisable as appropriate in a professional arena. In this they show that they can think in a 'designerly' way. The communication is primarily through visual means, and good drawing and modelling skills are very important. But it can be argued that demonstrating the ability to think creatively – and more particularly the creative synthesising of ideas and problems through design thinking – is the most important capability required to achieve this passport to enter the community of professional practice.

Recent research into design teaching has focused on its signature pedagogies – those elements which are particularly characteristic of the disciplines. Much of the most productive work has been based on core design theory, although this has often been enlivened by philosophies and approaches imported to the area. Most importantly such work has utility when it recognises the visual language of designing, the media of representation used, and the practical realities of tackling design questions. Increasingly the twenty-first century sees these activities in a global context where the international language of the visual artefact is recognised.

This book draws on recent work in these areas. It includes a number of chapters which are developed from work undertaken during the period of special funding for centres of teaching excellence in the UK up until 2010. Two of those in design have provided the basis for research and innovative developments reported on here. They have helped to enliven the environment for design pedagogy research in other establishments which are also included.

The Centres for Excellence in Teaching and Learning

Between 2005 and 2010 in England there was major funding for the development of teaching and learning in universities. The Centres for Excellence in Teaching and Learning (CETL) initiative represented the funding council's largest single funding initiative in pedagogy. It had two aims: to reward excellent teaching practice, and to further invest in that practice so that CETLs funding could deliver substantial benefits to students, teachers and institutions (HEFCE, 2011).

Seventy-four centres were funded across a range of universities, and within them a huge variety of types of pedagogic research and development was undertaken, across all discipline areas, much of it interdisciplinary and collaborative. Communities of Practice (CoP) figured quite noticeably within their range of activities, particularly in the area of professional development. A CoP was defined in that context as 'a group of people coming together from different disciplines or within a discipline for a common interest – pedagogical or subject-focused'. Sometimes these were formally organised within a discipline, and sometimes cross-faculty. It would seem that this type of arrangement would only loosely accord with Lave and Wenger's definition of a community of practice (Wenger 2007). However, it can be seen as evidence of the widespread currency of the notion within the initiative.

Across the 74 centres some 17 touched on 'creative arts and design' and thus may have been working in areas directly relevant to design pedagogy. Of course the many generic approaches which the centres engaged with may also have covered areas relevant to it. The number of centres which had a direct location in design schools was much smaller, and two of them covered work which focused directly on the development of practice-based education as a preparation for entry to the design profession. They were the Creative Learning in Practice (CLIP) CETL at the University of the Arts London, and the Centre of Excellence for Product and Automotive Design (CEPAD) CETL at Coventry University. CLIP had the specific aim to identify, evaluate and disseminate effective practice-based teaching and learning in the context of the creative industries. Similarly CEPAD was specifically orientated to facilitating the creation of portfolios which provided access to the community of international industrial design practice. Since 2010 staff who had been involved in those centres have carried on with developments in these areas.

Practice-based Teaching

Both CLIP and CEPAD operated in contexts where the pedagogy is predominantly studio-based. Traditionally art and design teaching is predicated on learning through doing, usually through the simulation of a professional situation by the means of a project brief. Students are neophyte designers engaged in the journey towards entering the community of professional practice of design.

The approach, which is typical of practiced-based design teaching, has a number of characteristics (Shreeve, Waring and Drew, 2008). Students are from the outset practitioners, often with long periods on projects, usually calling for a number of technical skills, and much activity is studio- and workshop-based. Assessment and feedback is usually through the 'crit' or 'critique' augmented by much peer learning. With less emphasis on formal knowledge there is acceptance of open-ended solutions, varieties of practice and tacit knowledge. Students are expected to become independent, self-analytical, critical thinkers, in an environment which does not emphasise theory, but does embrace key skills. Often a good proportion of the teaching staff are also practicing artists or designers.

Further Developments

This initiative to support a major investment in research and development for teaching and learning in English Universities with two centres where the pedagogy of design practice was a primary focus, served to embrace and utilise the idea of a community of practice as providing the arena for effective teaching and learning. This had particular resonance for the pedagogy of design practice with its natural emphasis on utilising members of the relevant professional communities within the teaching and learning arrangements. It also gave a focus to realising the explicit ambition of students of achieving the means to enter such communities of professional practice. These can be seen to require particular arrangements for studio teaching with partnership working. For some the crucial ability is to travel through an uncertainty threshold to achieve the transformative learning which is a key component in a community of practice. The legacy of these initiatives is not only the implementation of curriculum arrangements which embody these developments but also continuing research into the pedagogy of design practice.

One consequence was that a number of the staff who had worked in the centres formed the core of the Design Research Society's Special Interest Group in Design Pedagogy which was formed in 2009. The DRS has three main aims. It focuses on recognising design as a creative act, common to many disciplines. It has the intention of understanding research and its relationship with education and practice. Then there is the overall aim of advancing the theory and practice of design. The membership of DRS is international.

The Society's Special Interest Group in Design Pedagogy is one of five in the society. It aims to bring together design researchers, teachers and practitioners, and others responsible for the delivery of design education, and to clarify and develop the role of design research in providing the theoretical underpinning for design education. These aims are not directed simply at one type of design education, but are intended to include all ages. However, as the current membership of DRS is predominantly from universities, inevitably the research stream has concentrated on design education at that level. What is clear is that providing the research background to design education is one of the core purposes of DRS.

The Design Pedagogy SIG has now been able to collaborate with the CUMULUS organisation and others to bring into existence two research conferences in design pedagogy. Following the success of the first joint symposium for researchers into design education in Paris in May 2011, a second joint conference was held in Oslo from 14–17 May 2013. Its title was 'Design Learning for Tomorrow: Design Education from Kindergarten to PhD', and it was hosted by the School of Arts, Design and Architecture of Oslo and Akershus University College of Applied Sciences. With some 266 delegates and a very high standard of papers, the conference was a tremendous success. In these arenas the developments stemming from the CETLs have had wider international exposure and have contributed to the growing strength of design education research as a strand within design research overall.

Furthermore the chapters in this book are provided by members of that Special Interest Group. They are organised in two sections. The first is devised to set the scene, and the second to report on key developments in design pedagogy.

Part 1: Setting the Scene

CHAPTER 1: DESIGN EDUCATION RESEARCH: ITS CONTEXT, BACKGROUND AND APPROACHES

In this chapter Eddie Norman describes and discusses the development of research in design education through the designer, the design context and the design interface, and the background to developments in design pedagogy. He begins by discussing the problematic nature of design education research: the essential requirement for continuous curriculum development and the difficulties of making effective research contributions in this context. The characteristics and distinctions between design education in general and higher education are presented, and the consequential differences in research priorities discussed. Design in general education reflects conceptions of designing as a general human capability; and, in particular, the modes of thinking and learning that are made possible through designing, as opposed to the sciences and the humanities. This leads to a consideration of graphicacy, in contrast to numeracy and literacy, as a key human capability that makes designing possible. Design in higher education is characterised by the need to prepare students in particular design areas. Successful practice in different design areas requires the development of particular knowledge, skills and values, and hence a variety of course programmes and structures within higher education. The conception of design areas within the design field is presented and the differences in the associated knowledge skills and values illustrated.

He substantiates this position through an analysis of the theoretical positions underlying designerly approaches to research to this area, including case studies from the IDATER series of conferences, and its successors. From these he identifies three categories of design education research: the designer, the design context and the design interface, each of which provides a useful agenda for developing design education research.

CHAPTER 2: DESIGN EDUCATION AS THE PASSPORT TO PRACTICE

Central to the notion of a 'passport to design practice' is the recognition of the existence of groups as 'communities of design practitioners'. Where such communities are national and wear the badge of a professional body or society they are easy to identify and quite visible. However, there are other less formal international communities of design practice whose influence can be just as profound.

In this chapter Mike Tovey describes such communities within design as a related family. A powerful example is that of the community of practice of automotive designers. There are car design studios in all of the major industrial countries of the world, and the designers who work in them typically share their passion for automobiles and each time a new vehicle concept is revealed by one studio it causes interest and excitement in others. For an international community to function it is important that there is communication between its members. For automotive designers this is supplemented by online resources such as the Car Design News (CDN) website.

Students who wish to become proficient as designers devote their time to engaging with design project activity. This develops in intensity and detail and as students become more experienced they are able to tackle progressively more complex design problems. Typically the end goal is that of achieving a level of capability to function as designers in the professional world. That is, they wish to become part of the community of design practitioners. Today it is vital that their education is constructed so as to involve the design profession and mimic its practice.

CHAPTER 3: DESIGNERLY THINKING AND CREATIVITY

In this chapter Mike Tovey proposes a model of designing as a process which involves a peculiar and particular blend of thinking processes, which are the distinguishing characteristics shared by different sorts of designer. The designerly way of knowing makes use of various forms of intelligence, particularly visuospatial thinking. It is a peculiar and complex process which typically addresses those questions which are not precisely formulated and developed, or 'wicked problems'. Design thinking involves the use of parallel lines of thought deploying serial and simultaneous cognition. These can be corralled into two streams in a dual-processing model which aligns them with the preferred thinking styles of the two halves of the brain.

Such a model is consistent with a 'solution-led' approach and this is fundamental to its being a creative activity. It can encompass lateral thinking and diagonal thinking, which are examples of how such thinking skills can be taught and developed. At the core of the designerly way of knowing is a conversation between these modes of thought. Reflective practice is identified as an approach in which tacit knowledge can be deployed in reframing both the problem, and the solution. Various teaching strategies can accommodate these approaches. The studio, tutorial, library and crit are the traditional components, but using them effectively depends on the approach being informed by a deep understanding of the designerly way of knowing.

Part 2: Key Developments in Design Pedagogy

CHAPTER 4: FOSTERING MOTIVATION IN UNDERGRADUATE DESIGN EDUCATION

Designers are trained to deal with conflicting requirements and opportunities, and their ways of investigating problems and prototyping ideas are frequently aimed at exposing conflict to bring it out into the open. This requires a familiarity with a multidimensional landscape of design and designing.

In this chapter Steve Garner and Chris Evans discuss the role of student motivation in education. The nature of such motivation has been widely documented including the need to expose learners to challenge, risk and reward. However, there have been huge changes in learning and teaching and so fostering motivation is a very different challenge to even a decade ago.

Developing motivation in design education today presents some particular difficulties and opportunities. This chapter examines the vital stimulus of motivation in undergraduate product design education. It seeks to illuminate how students might develop their motivation through strategies such as developing curiosity, handling conflict, embracing failure and effective self-management that are sympathetic to design ideation and creative evaluation. In essence motivation supports effective creative and analytical thinking.

Those leading design education in universities place great emphasis on developing skills and knowledge, yet many expect students to automatically possess the necessary motivation for operating across today's design practice. Sometimes those who create design education assume their students must have the same drive and enthusiasm as themselves, while others assume that a hunger for success in the form of assignment grades or career opportunities is sufficient motivation. One might imagine that most designers are motivated by money, but the most powerful rewards in design are often those associated with being part of successful innovation, working as part of a team to successfully get a product into the marketplace where it's well received. It's here that undergraduate design courses can overlook such 'emotional' motivation and, even worse, create irrelevant reward systems.

Motivation is not an optional add-on. It has a function equal in significance to other intellectual and practical skills and knowledge. Motivation is not a vague, passive force; it can be understood, shaped and developed. In fact, designing demands a constant refreshing and renewal of motivation, and this has implications for design curricula and the sort of blended learning experiences that are created for students.

This chapter makes a case for prioritising the development of motivation in young designers. This is all the more urgent as forces conspire to erode motivation by swamping design tasks with information. Designers need support for the agile navigation of the world of design. We need learning experiences that tap into students' natural motivations but which professionalise motivation to create a resilient, informed and sustainable capacity. Since motivation is not one distinct force but is shaped and coloured by numerous cognitive forces and emotions, it seems logical that any attempt to develop motivation should acknowledge its diversity.

CHAPTER 5: SIGNATURE PEDAGOGIES IN DESIGN

In this chapter Alison Shreeve identifies the characteristics which are developed to prepare students for professions in design through engaging with the disciplinary ways of thinking, acting and being. In 2005 Lee Shulman proposed that learning in the professions had particular characteristics which enabled teachers to support their students to enter into the profession. He termed these 'signature pedagogies', ways to teach that enable people to develop disciplinary ways of thinking and being, or in other words, helping them to become certain kinds of professionals. For example, lawyers have their 'moot' and nurses use learning laboratories which simulate the hospital ward. This concept of learning has been more influential in North America, but the idea certainly has resonance with lecturers in higher education in the UK.

Based on a small research project at the University of the Arts London, Alison Shreeve and Ellen Sims identified particular ways to approach learning in art and design which they believed were signature pedagogies. These were distilled from the Landscapes project, which covered a range of disciplines. However, the principal of engaging others to learn to become part of a community of professional practice pretty much underlies how they approach teaching and learning in design. This chapter sets out the principle of signature pedagogies previously identified and explores what these are. It also identifies a number of ways in which tutors support their students to become designers which are more specific to particular disciplines, but help to illustrate the concept of signature pedagogies. This is a concept which merits further more detailed investigation at the individual design subject level.

CHAPTER 6: THE EXPERIENCE OF TEACHING A CREATIVE PRACTICE

In this chapter Linda Drew explores conceptions of teaching held by academics in departments of design and explores links between those conceptions and the communities of practice associated with the subject context. She explores the qualitatively different ways that teachers of design experience their teaching. The study focuses on teachers of practice-based subjects in design. Much of the work that has examined teachers' conceptions has built on research frameworks that also explored students' conceptions and approaches to learning. Studies of conceptions of teaching have ranged from the phenomenographic to those studies of belief orientations. The data is from an interview study of eighteen teachers from eight UK universities and is explored with a phenomenographic approach. The analysis identified the qualitatively different ways teachers of design experience their teaching within which variation in the practice dimensions could also be discerned.

The chapter reports variation between the qualitatively different ways the teachers conceive of teaching. The important feature of this analysis is the community of practice dimension, in particular how teaching is perceived as contributing to engaging with the social practices which constitute the particular design practice. The community of practice dimension is further explored in relation to how teachers may enhance the experience of learning and the learning environment by developing strategies which address the application of knowledge in practice-based settings as well as their activity systems. Participation in a community of practice is a key premise to understanding learning to practice, including learning the values and appropriating an identity related to that practice.

CHAPTER 7: TRANSFORMATIVE PRACTICE AS A LEARNING APPROACH FOR INDUSTRIAL DESIGNERS

In this chapter Karen Bull outlines an approach to teaching design practice through transformative learning. Becoming a successful designer depends strongly on individual capability to think in a designerly way, as well as the specialist design skills and knowledge to translate and develop ideas. Core to this is a process that requires the integration of both holistic and linear ways of thinking in a dual-processing model, through engaging in practice. Typically this involves design projects, experiential problem-solving and creative experimentation. This is not an approach to thinking that is easy to achieve within a course structure based on the delivery of modular packets of teaching which emphasise efficiency and accountability.

The revised approach aims to reinforce the benefits of a transformative practice-based approach to learning. It has been developed from research undertaken through Coventry University's Centre of Excellence in Product and Automotive Design (CEPAD) and is based on a formal longitudinal study of student experience and other related case-study activity. Central to the approach is the development of mechanisms which help students to surmount a threshold of uncertainty. This is typically connected to the troublesome experiences that are encountered when engaging with design problems and solutioning. By embedding the activity in a studio-based culture of learning that engages students through practice and experiential learning it can be handled effectively and productively. What is required is a learning environment which overcomes the fragmentation that can occur on modular curriculum structures and provides most significantly 'safe spaces' for problem-based learning. These allow students to take creative risks without the fear of failure. A course organisation based on 'scaffolding' structures provides the space for students engaging in learning which takes them from from aspiration to a strong personal alignment with a professional community of design practice. By using a Gateway Assessment Model for evaluation and feedback it can be embedded into a curriculum that incorporates collaborative and peer-based learning. Within this arrangement tutors can work with students to define individual learning pathways. The experiential learning is reinforced by incorporating inputs focusing on global industry awareness and a variety of industry inputs to the curriculum such as collaborative projects and industrial placements. These elements all serve to reinforce student aspiration, confidence and professional awareness through a transformative practice-based model of learning. The model supports students as they develop from novices aspiring to be a designer to becoming confidently 'ready' to enter their professional community of design practice.

CHAPTER 8: INDUSTRIAL DESIGN AND LIMINAL SPACES

In this chapter Jane Osmond reports on research, using the threshold concept theory developed by Meyer and Land. She followed a cohort of undergraduates from entry to graduation between 2005 and 2010: in total 89 students were interviewed during the lifetime of the research. The threshold concept theory posits the idea that within disciplines there are conceptual gateways or portals, which – due to their troublesome nature – can make it difficult for students to progress. This notion of a threshold concept is seen as distinct from 'core concepts' – or building blocks – within disciplines, as it engages with the notion of transformation. Grasping, experiencing and understanding a threshold concept will irrevocably transform a student's understanding, and this transformation can relate to the particular subject at hand, and/or be extrapolated beyond the academy.

In this chapter she outlines the importance of the concept of liminal spaces in relation to a creative curriculum within higher education. Based on the work of the Centre of Excellence for Product and Automotive Design (CEPAD), which identified the toleration of design uncertainty as a threshold concept, there is a consideration of the notion of liminal spaces in relation to both the creativity and design literature. It is argued that for student designers, liminal spaces can be unsafe places as they will not have the skills, experiences and confidence necessary to negotiate them successfully and so a curriculum, which first identifies its 'jewels' and second builds safe spaces surrounding them, can only enhance students' creative abilities. In conclusion, there seems little doubt that giving students the time, space and structure to immerse themselves into a design brief can only enhance their creativity and design solutioning ability.

CHAPTER 9: DEVELOPING TOOLS TO SUPPORT AND UNDERSTANDING DURING INDUSTRIAL DESIGN PRACTICE

In this chapter Mark Evans, Eujin Pei and Ian Campbell consider various aspects of visual design representation. Both design thinking and design communication make use of various forms of representation including sketches, drawings, models and prototypes which can be drawn together as a taxonomy indicating their varied functions. The use of case-study tools to support design education, a range of tools from 3D modelling to colour specification which have been validated through use by designers in professional practice, are introduced to design-studio teaching.

The chapter explores the barriers to communication during product development with a focus on industrial designers and engineering designers. It provides a thoroughly referenced indication of why difficulties occur and develops a strategy to help resolve language issues and problems in understanding through a knowledge framework for design representations. The research was undertaken in four phases. Barriers to communication were identified through semi-structured interviews with industrial designers and engineering designers at a number of industrial design consultancies. The nature of design representations was categorised (as sketches, drawings, models or prototypes) using interviews with both industrial designers and engineering designers, and the differences in use between the two groups identified. By the use of a process of information design to translate the findings and data from Phase 2 into the card-based CoLab design tool that included the taxonomy it was possible to identify when the design representations were used by industrial designers and engineering designers and for what types of information. Subsequently the final tool was validated through further interviews with industrial design and engineering design practitioners and observation. The final phase involved disseminating the research output with the support of the Royal Academy of Engineering (RAE) in the UK (CoLab web-based design tool) and Industrial Designers Society of America (IDSA) in the USA (iD Cards physical design tool).

This approach has the potential to elicit valuable and unexpected tacit knowledge that can contribute to student learning. It also acknowledges that while the outcomes from such research can be enthusiastically received, translation into a format for effective dissemination can be a challenging and time-consuming process. However, with confidence in outcomes and a desire to disseminate, opportunities can be identified. There is considerable scope for educators to integrate these resources into their teaching as required. It is a tool which illustrates both the disciplinary differences between the two communities of professional practice of industrial design and engineering design, and provides a basis for teaching strategies to deal with and capitalise on those characteristics.

CHAPTER 10: THE USE OF DESIGN CASE STUDIES IN DESIGN EDUCATION

In this chapter Seymour Roworth-Stokes and Tim Ball explore the appropriateness and relevance of case studies to design education and their use in the teaching of design practice. Case studies provide a rich insight into design proposals which are often grounded in real life and complex situations. It is argued that they offer the potential to understand design methods through good design practice when there is a clear context underpinned by sound empirical evidence.

The chapter commences with the findings of an audit of 223 design case studies in 4 of the leading design research journals with a detailed analysis of the type, subject and field of research identified within the leading journal *Design Studies* since 1979. This is used to highlight the range and subject coverage of available case studies as a precursor to a detailed exploration of the use of case study in teaching. Drawing upon a practice-based industrial design department in a UK university with long-established industrial links and partnerships, examples are provided to describe how case studies are employed within the curriculum to identify and draw out design methods and approaches. Reflections of staff and students consider the notion and relevance of 'cases' to inform a predominantly studio-based learning environment.

The chapter concludes with guidance on the use of case studies in design education and a discussion concerning their role in the development and acquisition of advanced, and professionally relevant, design skills and competences.

CHAPTER 11: AMPLIFYING LEARNERS' VOICES THROUGH THE GLOBAL STUDIO

In this chapter Erik Bohemia and Aysar Ghassan argue that understanding the construction of autobiographical processes is an important aspect of gaining entry to the community of professional practice. A central part of constructing such a narrative is learning how to tell appropriate stories about oneself to prospective employers. Consequently it is argued that design students must learn to tell their own stories. The commonly utilised master–apprentice model which is highly useful in areas such as skills acquisition it may not be optimally effective in aiding students to tell their own stories. Other approaches are required if we are to enable future design graduates the necessary reflexivity to be able to negotiate the increasingly complex world of the contemporary knowledge economy.

The Global Studio aims to propagate a student-led pedagogic model in which tutors purposefully try to maintain their distance so as to encourage autonomy. The aim is to introduce learners to 'complex project situations' and consequently to prepare them for contemporary working life. It is operationally different from 'tutor-led' design education as lecturers are more 'distant' in teaching and learning activities. Students construct conversations and outcomes primarily via interaction with peers.

Feedback from home-institution students suggests many individuals struggle with making decisions without tutor-led involvement from tutors. The global studio moves on from design education in which lecturers are at the centre of teaching and learning activities and where educators' tastes strongly influence students' outcomes. The intention is to prepare graduates for working in highly complex professional capacities synonymous with the contemporary era in which interactions with their peers are key to creating self-confident design graduates attuned to the contemporary community of design practice. As tutor-led design education is inherently resource-intensive, the model offered by the Global Studio may be a more effective and affordable way forward.

CHAPTER 12: CONCLUSIONS

The main conclusion offered from this collection of pieces by members of the Design Pedagogy group is that this area of design research is essential for developing our understanding of design practice. This in its turn is necessary if degree-level design education is to be delivered effectively and well.

Designing is a peculiar process with its own culture and atypical thinking processes. It is also an international activity of great economic significance. Consequently undertaking research to inform and better deliver design graduates fit to enter the community of professional design practice is a worthy enterprise.

Michael Tovey

References

HEFCE (2011) *Summative Evaluation of the CETL Programme: Final Report by SQW to HEFCE and DEL HEFCE*. HEFCE, Gloucester.

Shreeve, A., Wareing, S. and Drew, L. (2008) Key aspects of teaching and learning in the visual arts, in H. Fry, S. Ketteridge and S. Marshall (eds), *A Handbook of Learning in Higher Education*, 3rd edn. London: Kogan Page.

Wenger, E. (2007) *Communities of Practice Learning, Meaning and Identity*. Cambridge: Cambridge University Press.

PART I
SETTING THE SCENE

Chapter 1

Design Education Research: Its Context, Background and Approaches

EDDIE NORMAN

Design education research is not a new area of activity, and there is a plausible case for considering its origins in the work of Pestalozzi (1746–1827), Fröbel (1781–1852), Cygnaeus (1810–1888) and Salomon (1849–1907), who developed the Sloyd approach (see Ólafsson and Thorsteinsson, 2009). Such early authors, and some more recent ones have developed positions in regard to aspects of design education, implemented them and described the outcomes. Such 'curriculum development' is not always recognised as 'research' with the underlying concern being that the outcomes of such work are context-specific, and consequently difficult or impossible to transfer or generalise. It is commonly argued that such universal truths are the appropriate goal for genuine research, and it is the credibility of this position in the context of design education that is the central concern of this chapter.

The chapter begins by considering the difficulties in developing generalised models that could frame research investigations and then moves on to discuss the strategies that design researchers have adopted over the past few decades in order to make effective research contributions. These can be classified as studies of the designer, the design context and the design interface, and these classifications provide a background from which to understand developments in design pedagogy.

Towards Generalised Models for Design Education Research

Whether explicitly or implicitly design education researchers position themselves in relation to two key factors, namely:

- the relationship of general and higher education;

- the boundaries of their research.

The issues surrounding these two areas are central to understanding the problematic nature of design education research, and consequently the approaches that are most likely to lead to effective research contributions.

GENERAL AND HIGHER EDUCATION

If this book was about the education (training) of Olympic athletes, or indeed elite performers in any sports, then it would almost be taken for granted that the roots of the task must lie in the formative years of general education. There would be debates concerning the appropriate nature of the early sporting experiences, but an expectation that it would be from these that the elite would be identified and within these that they would initially be nurtured. This is because sporting capability is conceived as something that everyone possesses to some degree, as well, of course, as something that specialist training can develop. Design education emerged into the mainstream of general education in the UK in the 1970s and 1980s, and largely as a result of the work of the Design Education Unit (DEU) at the Royal College of Art led by the late Professor Bruce Archer. At the DEU a dual model of design capability as both universal and specialist emerged, and, although much has changed since the 1980s, this is one position that many of us believe has stood the test of time. Consider this 2006 passage from Ken Baynes who worked with Bruce Archer, and later led the DEU, reflecting on this dual model.

> 'These developments in the socio-political realm have been complemented by extraordinary progress in cognitive science and neuroscience. Work in these fields throws new light on the nature of 'designerly' thinking and on the personal, social and economic importance of these thought processes.
>
> What cognitive science has done is to show conclusively that designerly thinking and action are features of the mental activities of all humans.

It has settled the argument between two apparently contradictory views of design.

1. *That design is highly specialist, complex and esoteric – that particularly the act of designing is something which people can do only after a long apprenticeship.*

2. *That design ability, like language ability, is something that everyone possesses at least to some degree.*

We now have to accept that these two views are in fact complementary. The highly complex skills of the professional engineer, fashion designer or CGI artist are simply the specialist development of abilities and understandings that we all have.

The design education 'movement' always took the broad view and in doing this they were building on a distinguished tradition that included William Morris, W R Lethaby and Eric Gill.' (Baynes, 2006: 7)

Detailed discussion of the evidence for this position can be found in Ken Baynes' book *Design: Models of Change* (2013), and the key point to note here is that not everyone agrees. Some design education researchers in higher education do not position themselves at the end of food chain over which they have little or no influence, but see their research as something that can be evaluated independently of this complexity.

Design education researchers in general education would naturally acknowledge that their students are developing, and their capabilities changing, as a result of the influence of 'nature' alongside 'nurture'. (Nature's influences here have to be interpreted to include those from cultural as well as biological sources.) Again some design education researchers in higher education – and I would suggest most – tend to assume that their students have reached 'maturity' and that changes in design capability following teaching interventions can be attributed solely to nurture. Recent research on the human brain has shown that development can continue into 'adulthood', so this position is also at risk.

BOUNDARIES

There are occasional, perhaps understandable suggestions that education research should be more like medical research – large trials and placebos presumably. Leaving aside the question of what a 'placebo education intervention' might comprise (presumably repeating a previous programme while convincing staff and students that it was 'new', although it is hard to imagine), the design of such research requires knowledge of all the variables (factors, degrees of freedom) involved. This would certainly require detailed analysis and is by no means straightforward. Indeed, there must be some doubt as to whether or not it is actually achievable. Consider the following example.

Alexandros Mettas is a teacher and teacher educator in Cyprus and for his Ph.D. research he set out to better understand the design decision-making capabilities of his pupils aged 12–15. He conducted a detailed literature review, a pilot study and a main study involving pre-tests, post-tests, observations of design tasks and interviews with the pupils, and analysed these using a grounded theory approach. One of the outcomes was the model of the factors affecting design decision-making for his pupils (shown in Figure 1.1). At the centre were the requirements of the Design and Technology curriculum in Cyprus. There were the factors that relate to any age group – their knowledge, skills and values. There were factors relating to the teacher and the teaching resources, and there were particular factors relating to the children, such as the lack of transfer between school activities and other areas of their lives. In the current context it is not the model itself that is significant, but the way it illustrates the complexity of the design education research context. In order to isolate the effect of one particular factor, research design would need to take account of all these factors.

For example, one aspect of generalising the research findings would be to establish whether the same model also applied in other countries. Mettas' literature review explored research conducted in many different countries, but the remainder of the research was conducted in Cyprus, and while meeting the requirements of a particular national curriculum. Perhaps further, and even more dominant factors, might appear if the research was repeated with different designing tasks, teachers and in another country. As they read this, design education researchers working with other age ranges and in different design areas are probably already noting differences that relate to their context. So how realistic is it to expect research questions relating to design education to be fully defined, or are they showing the characteristics of 'wicked' (or ill-defined) problems with which designers are all too familiar?

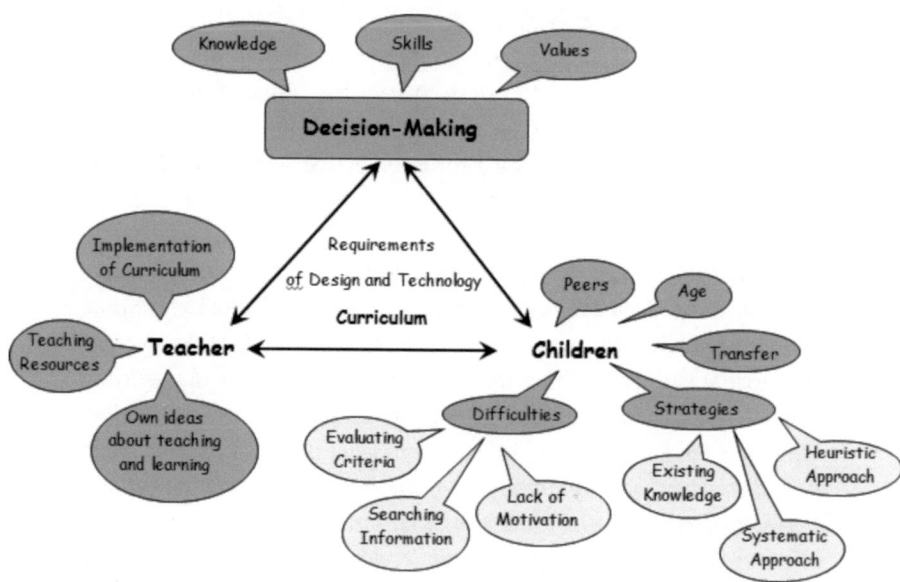

Figure 1.1 **Model for factors affecting children's decision-making in design and technology (Mettas, 2012: 188)**

Design education researchers approach these difficulties through essentially two strategies. First, they impose boundaries on their research by specifying age ranges, particular design areas and particular contexts. The most limited research contributions apply only to the context from which they were derived: 'this was the initiative with my students and these were their results'. The more sophisticated contributions will define and analyse the boundaries of the research study in order to maximise the possibilities of transfer. Second, they use 'designerly' research methods which are discussed below.

Designerly Research Methods

Significant efforts were made to address the conceptual issues surrounding the question of how design education research could be best approached during the 1990s, and particularly at *IDATER* (the *International Conference on Design and Technology Educational Research and Curriculum Development*) which ran from 1988 to 2001. The *IDATER* conferences were established to support the development of a research base for the introduction of the national curriculum in Design and Technology in England in 1990, but rapidly grew towards greater significance. They became international in 1992 – *DATER* became *IDATER* – and there were strands of papers organised by the Design Research Society (DRS) at *IDATER99* and *IDATER2000*. The conferences always sought to support practitioner research, as illustrated by two key theoretical contributions to the understanding of action research as a designerly mode of enquiry. These were made in the Keynote Addresses by Professor Bruce Archer at *IDATER91* and Professor Phil Roberts at *IDATER2000*.

A designerly approach, rather than a scholarly or scientific approach, can with advantage be made towards educational research and curriculum development. Design, in a certain sense, is research done backwards. Research starts with the particular, and moves towards the general. Design starts with the general and works towards the particular. Designers are told, or decide, at the outset, what their end product must be and do. They begin by conceiving of one or more broad configurations that seem likely to be, and to do, what is required. They then elaborate the structure of these configurations and develop the subsystems of one or more of the most promising proposals. They then detail the construction, working backwards to the particular, the bits and pieces, upon whose correct construction depends the efficacy of the whole. At various stages, the validity of assumptions is checked and performances are measured. The same basic design process can be, and is being, applied to the development of all sorts of artefacts and systems that have not hitherto been thought of as subjects for design. For example, providers of banking and other financial services now speak of their products (that is, charge cards, insurance policies, etc.) as having been designed to meet the needs of given classes of the user. Curricula, courses, lessons and examinations are thus proper subjects for design. Happily, the National Curriculum Council's attainment targets provide ready-made design requirement specifications. A designerly approach to curriculum or course design might be to ask:

What sort of capability profile would a pupil need to exhibit in order to be seen to have attained the target in question? And then:

What are the categories of knowledge, skill and values that contribute to such a profile?

What are the components of each category?

What kinds of learning experience are likely to imprint each of these components of knowledge, skill and value?

How can such learning experiences be provided?

and so on, from the general to the particular, down to exercise design, performance assessment design and resource allocation. There is every reason for teachers of design and technology to use the techniques with which they are familiar to attain the objectives to which they are committed.

I opened this address with the question:

What kind of research is appropriate to the study of education through Design and Technology?

My answer has been:

The designerly mode of enquiry is entirely appropriate to the study of education through Design and Technology. It is also less prone than are scholarly or scientific modes of enquiry to distortions arising from conflicts. (Archer, 1992: 12)

While attracting support from design education researchers, the designerly strategy that Bruce Archer had proposed had remained controversial, which was one of the reasons that it was revisited by Phil Roberts. Among the objectives of Roberts' Keynote Address were the support of action research as a mode of inquiry and development that is especially appropriate to D&T educational practitioners; the support of the teacher-as-researcher (or practitioner-as-researcher); and the support of the position that action research within education (and D&T education) is intended to improve practice. He described action research as follows.

At its simplest, classroom action research relates to any teacher who is concerned with his/her own teaching: to the teacher who is prepared to question his/her own approaches in order to improve the quality of teaching and learning. Hence, the teacher/practitioner is involved in looking at what is actually going on in the classroom [or studio/ workshop]. He/she seeks to improve his/her own understanding of a particular problem (or state of affairs) rather than to impose an instant 'solution' upon that unarticulated problem. It is crucial that time be taken for thought and reflection, and it is implicit in the idea of action research that there should be some practical effect of, or end product to, the research which would be based on a now increased awareness of what actually happens in the classroom. It is, as a consequence, towards the construction of a practitioners' theory, constructed from their experience; and it would intend to be useful.

On this view, some of the characteristics of educational action research are that:

1. *its activities and objects are concerned with the deepening of understanding of the studio, workshop, classroom, and school situation by the teacher/researcher adopting a critical, questioning stance. Its starting points are the 'practical problems' experienced by teachers, rather than the problems found within the formal theories of the 'education disciplines'.*

2. *The presentation of its reporting is in ordinary everyday language, and might well take the form of a case study or story. It adopts the action perspective of practitioners and employs their everyday language to describe and investigate its subject-matter states of affairs.*

3. *Reflection on experience is part of its processes.*

Not all would agree with this, obviously simplified, characterisation of action research, and one of IDATER's functions should be to stimulate discussion about its nature and nuances. (Roberts, 2000: 18)

These contributions by Archer and Roberts were recognising that research problems in design education are real world problems and hence ill-defined (wicked) problems. Unless they are to be abstracted from their context, then it is designerly methods that are most likely to be effective.

Effective Design Education Research Contributions

From the preceding discussions it can be suggested that effective contributions to design education research will have the following characteristics:

*   the position being taken concerning the emergence of design capability as a general and/or specialist aspect of human development will be clear;

- the relationship of nature and nuture in relation to the interpretation of outcomes will be considered;

- the boundaries of the context, student population and design areas being researched will be analysed;

- and designerly research methods may have been employed.

In an effort to go beyond this position, the author presented evidence relating to the analysis of two particular case studies: the *IDATER* (*International Design and Technology Education Research and Curriculum Development*, 1988–2001) and the subsequent *Design & Technology (D&T) Association Education and International Research Conferences* (2002–2010) at the *First DRS/Cumulus Conference* (Norman, 2011). After the last of the *IDATER* conferences, the D&T Association conferences continued its efforts and traditions, and also brought the research contributions into closer proximity with teachers.

This analysis considered the originators of the research, the methods they used and their intentions. It turned out that useful categories within which to consider the nature of contributions to design education research comprised:

- the designer;

- the design context; and

- the design interface

CATEGORIZING RESEARCH CONTRIBUTIONS

Research contributions could be analysed as intending to target one of three areas.

- The designer(s): the individual(s) their capabilities and their competences for designing.

- The design context: the analysis of the knowledge, skills and values that they might possess.

- The interface: tools for designing and organisational structures that enhance designer's capabilities, competences and access to their context.

Some discussion of each of these is necessary to explain their use.

The Designer

Human capabilities are characteristics that can be developed and people are said to be competent when they have sufficient knowledge, skills or values for a particular purpose. The relationships indicated in Figure 1.2 can be understood by asking:

- What makes using the imagination and the senses possible? One answer to this question would be modelling (Baynes, 2009a); a capability in which would appear to be a fundamental human characteristic (Doyle, 2004). Modelling embodies the use of the imagination and senses in the context of designing.

- What makes modelling possible? Developed competences in numeracy, literacy, articulacy, and graphicacy clearly play their part, and are what is immediately evident in an educational context. However, this question runs much deeper into areas of cognitive psychology.

- And, for example, what makes graphicacy possible? Developing competences in areas such as perspective would again be part of the unfolding story, but these will only be possibilities as competence in areas such as mark-making emerge (Danos, 2012).

The ways in which these capabilities and competences find expression change. The use of the imagination and the senses is interwoven with the development of human societies and culture; modelling strategies are linked to economic development (Baynes, 2009b) and technological change, and developed and emerging competences are linked to the consequences of such development. For example, the place of drawing in children's upbringing and early approaches to numeracy are linked to the evolution of electronic products and communication technologies. It is commonly recognised that humans alter and influence their surroundings, but there is less frequent recognition that such changes then go on to alter human development. Humans are fully situated in their contexts.

Human capabilities ... including the use of the
imagination and the senses

Modelling ... alongside composing, writing ...

Developed competences ... numeracy,
literacy, articulacy, graphicacy ...

Developing competences ... manipulation,
vocabulary, expression, perspective ...

Emerging competences ... logical reasoning,
language, mark making ...

Figure 1.2   **Human capabilities and a hierarchy of competences in the
context of designing (Norman, 2011: 58)**

So, in relation to this consideration of the designer's capabilities and competences the interdisciplinary nature of design research is apparent, and the associated potential for contributions from many areas. The need for continuous reappraisal of these matters and associated curriculum development is also evident.

The Design Context

Analysing the knowledge, skills and values that relate to the design field or particular design areas is another possible strategy for seeking to contribute to design education research. There is, of course, no reason to assume correlation between the knowledge, skills and values that designers in a particular design area could possess and those that they do possess, and the gaps help to define targets for curriculum planners and policy-makers, for the designers of 'tools for designing', and for continuous personal and professional development. It is possible to research these areas separately, e.g. knowledge (de Vries, 2003; Friedman, 2001), skills (Design Skills Advisory Panel, 2007) and values (Trimingham, 2007). It is also possible to research them together under headings such as sustainable design and design for emotion.

When considering design education research relating to 'the design context', the inevitable overlaps with design research become apparent.

THE INTERFACE

Design tools, such as computer-aided design and manufacture (CADCAM), the Cambridge Engineering Selector (CES) materials database, computer-aided ergonomic modelling (e.g. SAMMIE CAD[1]), Ardhuino[2] and web-based design guides (e.g. *Information-Inspiration*[3]) change the relationship between the designer(s) and their context. They provide access to 'capabilities and competences' that can far exceed those the designer(s) can possess without such technological enhancement.

If you accept a wide enough definition of technology that includes social and economic organisation, then sufficient has already been said, but it is worth noting that team or group work, a well-designed and resourced working environment, social networking and well-managed supply chains can also enhance designer's capabilities, and, of course, those of design students. Hence the management and organisation of design education are essentially concerned with the interface, whereas its aims are related to the evolving design context.

So with effective design education research contributions spanning the designer's changing capabilities and competences, the evolving contexts of their designing, together with tools and organisational structures developed to support their interface, the complexity of the research area is apparent.

ANALYSING THE IDATER AND D&T ASSOCIATION CONFERENCES AS CASE STUDIES

Four conferences at five-yearly intervals were selected for analysis as shown below. These case studies represented this series of conferences at their high points and avoided including issues relating to getting started or transferring the conference's governance in the data analysis

---

1   For information about SAMMIE CAD see http://www.sammiecad.com/
2   For information about Arhuino, which is an open source electronic prototyping platform, see http://www.arduino.cc/
3   For information about *Information-Inspiration* see http://www.informationinspiration.org.uk/

- *DATER90*: two years after the first conference and when it was still finding its feet

- *IDATER95*: at the peak of its influence

- *IDATER2000*: towards the end and the year in which the decision was taken to run 'just one more'

- *ID&TA2005*: held at Sheffield Hallam University and one of the more influential of the new conference series

The results for the analysis of the research outputs reported at these conferences are shown below in terms of their originators (Table 1.1), research methods (Table 1.2) and intentions (Table 1.3).

In relation to Table 1.1, *IDATER* might be considered to be at its most successful at the point at which the decision was made to end the series (i.e. in 2000). The number of authors of research outputs from England had reduced to 14, but 43 of the 48 authors were from higher education. Clearly the links to, and impacts on, practice in general education were at risk. However, disappointingly, the move to organise the research conference alongside the education conference at the D&T Association in 2005 did not significantly alter this position, except in increasing the proportion of contributors from England.

Table 1.2 shows that the nature of the research methods employed was changing. At *DATER90*, the dominant form of activity was document analysis; at *IDATER95* the use of empirical data and case studies to support the document analysis emerged; at *IDATER 2000* case studies had become the major research activity; and at *ID&TA2005* empirical data and its use in support of case studies were central.

Table 1.3 shows how the research intentions of the contributors switched over this 15-year period. At each conference there were research outputs focussing on the designer and the development of their capabilities. However, there was a clear movement away from outputs related to the design context and towards the interface between the designer and their context.

Table 1.1   **Originators of contributions to IDATER and D&T Association Conferences[4] and the 2011 DRS/Cumulus Conference**

		1990 DATER	1995 IDATER	2000 IDATER	2005 ID&TA	2011 CUMULUS/ DRS
Total number of papers		28	32	26	21	16
Total number of authors		37	43	48	34	18
Affiliations of authors	*Higher Education (academic)*	21	37	41	31	17
	Higher Education (research assistant/ student)	2	1	2	2	
	General education	6	1	3	0	
	Education other	5	3	0	1	
	Design/ Industry	3	1	2	0	1
Countries of origin		England (× 36) USA	England (× 29) Australia (× 4) Botswana Bulgaria Germany Greece Hungary Scotland (× 2) USA (× 2) Zimbabwe	England (× 14) Australia (× 2) Canada (× 3) Indonesia Israel (× 2) New Zealand (× 2) Northern Ireland (× 3) Taiwan Wales (× 8) Japan (× 3)	England (× 25) Australia (× 3) Canada (× 2) Cyprus (× 2) Northern Ireland Sweden	England (× 8) Finland (× 1) France (× 5) Italy (× 2) Netherlands (× 1) Switzerland (× 1)

---

4   Keynote speakers have been excluded and a maximum of three authors per paper recorded in order to avoid distorting the analysis.

Table 1.2   Research approaches of contributors to IDATER and D&T Association Conferences and the 2011 DRS/Cumulus Conference

	1990 DATER	1995 IDATER	2000 IDATER	2005 ID&TA	2011 CUMULUS/ DRS
Document analysis	12	9	2	1	4
Document analysis + case studies	4	7	5	0	9
Document analysis + empirical data	2	13	3	4	3
Case studies	5	0	9	0	
Empirical data	4	3	2	6	
Case study + empirical data	0	0	3	12	
Literature review	1	0	0	0	
Literature review + empirical data	0	0	2	0	

Table 1.3   Research intentions of contributors to IDATER and D&T Association Conferences and the 2011 DRS/Cumulus Conference

	1990 DATER	1995 IDATER	2000 IDATER	2005 ID&TA	2011 CUMULUS/DRS
The designer	7	10	5	7	0
The design context	13	10	9	2	4
The interface	8	12	12	12	12

REFLECTION ON THE DRS/CUMULUS CONFERENCE IN 2011

The original versions of Tables 1.1–1.3 were created to explore a particular research question in support of the First International Symposium for Design Education Researchers organised by Cumulus/DRS SIG in Paris in May 2011 (Norman, 2011). The research question which had seemed appropriate for this important event was:

- *Are effective contributions in design education research significantly different to effective research contributions in general education?*

For the purposes of this chapter, the papers submitted to the First Cumulus/DRS conference have been analysed and placed in an additional column. The sixteen papers could be classified under the three headings of the designer, the design context and the interface, which is both important in supporting the robustness of the framework and revealing in demonstrating the consistency of the approaches used by researchers in general and higher education. The analysis also revealed few attempts at providing supporting empirical evidence and no research studies of 'the designer'. The latter is not too surprising and suggests the predominance of the 'mature adults' perspective, but the lack of empirical evidence is a little unexpected.

It can be observed that most authors at the First Cumulus/DRS conference did not offer empirical evidence in support of the positions that they presented, but typically they did offer case studies in particular contexts and some analysis of the boundaries. It is for other researchers to discuss their approaches: I will confine my comments here to the paper that I contributed. It has always been a cause of some amusement to me during my research career to hear people refer to me as a 'positivist', as if I needed some kind of external evidence for everything. However, it is not really positivism that motivates my searches for some kind of corroborating data, but fear. Amongst the subjects I taught during my career was mechanics, and those familiar with this area will be aware of Aristotle's law of motion (essentially, 'if a body is moving, then there must be a force acting'). This held sway for around 2000 years and it was only Newton's observations of planetary motion that revealed the error, i.e. that forces are only acting when the speed or direction of motion is changing. However good the philosopher, just looking at a real world issue and theorising is full of risks. Gathering empirical evidence does not remove the risks, after all it is often just another 'white swan', but it can reduce them a little.

The Second Cumulus/DRS conference took place in Oslo in May 2013 and there were around 150 papers submitted. It would be expected that as this conference matured the use of empirical data and detailed literature reviews towards the causes of generality and transfer would appear more often, but no analysis has been undertaken for this book chapter. The response to the call was magnificent and demonstrated the growing importance of design education research. The contribution of the Oslo conference deserves more careful consideration than space here permits.

The paper for the Paris Symposium from which this chapter was developed was looking at design education research as a strand of education research, and took a particular approach in order to explore some of their similarities. This chapter has been focused on revealing some of the distinctive characteristics of design education research as background to the development of design pedagogy.

References

Archer, L.B. (1992) The nature of research into design and design education, in B. Archer, K. Baynes and P.H. Roberts (eds), *The Nature of Research into Design and Technology Education*, pp. 7–13. Loughborough: Department of Design and Technology, Loughborough University of Technology. http://magpie.lboro. ac.uk/dspace/bitstream/2134/1687/1/Archer_Baynes_Roberts.pdf.

Baynes, K. (2006) Design education: what's the point?, *Design and Technology Education: An International Journal*, 11(3), 7–10. http://ojs.lboro.ac.uk/ojs/ index.php/DATE/article/view/Journal_11.3_1006_REF/74.

Baynes, K. (2009a) *Models of Change: the Impact of 'Designerly Thinking' on People's Lives and the Environment: Modelling and Design*. Loughborough: Department of Design and Technology, Loughborough University. https://dspace.lboro. ac.uk/dspace-jspui/handle/2134/5165.

Baynes, K. (2009b) *Models of Change: the Impact of 'Designerly Thinking' on People's Lives and the Environment: Modelling and Society*. Loughborough: Department of Design and Technology, Loughborough University. https://dspace.lboro. ac.uk/dspace-jspui/handle/2134/6092.

Baynes, K. (2013) *Design: Models of Change*. Shepshed, UK: Loughborough Design Press.

Danos, X. (2012) *Graphicacy Within the Secondary School Curriculum, an Exploration of Continuity and Progression of Graphicacy in Children Aged 11–15.* Ph.D. Thesis, Loughborough University. https://dspace.lboro.ac.uk/dspace-jspui/handle/2134/9652.

Design Skills Advisory Panel (2007) *High-level Skills for Higher Value.* Design Council & Creative Cultural Skills, http://www.ukdesignskills.com, Design Council, London.

de Vries, M. (2003) Design matters, and so does philosophy of design: John Eggleston Memorial Lecture, *Journal of Design and Technology Education*, 8(3), 150–153. https://ojs.lboro.ac.uk/ojs/index.php/JDTE/article/view/648.

Doyle, M. (2004) The evolution of technicity: whence creativity and innovation?, in E.W.L. Norman, D. Spendlove, P. Grover and A. Mitchell (eds), *Creativity and Innovation: DATA International Research Conference 2004*, pp. 67–72. Wellesbourne: The Design and Technology Association (DATA), Wellesbourne. https://dspace.lboro.ac.uk/dspace-jspui/handle/2134/2837.

Friedman, K. (2001) Creating design knowledge: from research into practice, in E.W.L. Norman and P.H. Roberts (eds), *Design and Technology Educational Research and Curriculum Development: The Emerging Research Agenda*, Loughborough: Department of Design and Technology, Loughborough University. https://dspace.lboro.ac.uk/dspace-jspui/handle/2134/953.

Ólafsson, B. and Thorsteinsson, G. (2009) Design and craft education in Iceland, pedagogical background and development: a literature review, *Design and Technology Education: An International Journal*, 14(3), 10–24. http://ojs.lboro.ac.uk/ojs/index.php/DATE/article/view/246.

Mettas, A. (2012) *Design Decision-making by Children Aged 12–15 Within Design and Technology Education.* Ph.D. Thesis, Loughborough University. https://dspace.lboro.ac.uk/dspace-jspui/handle/2134/9251.

Norman, E. (2011) The nature of effective research contributions in design education, in E. Bohemia, B. Borja de Mozota and L. Collina (eds), *Researching Design Education: Ist International Symposium for Design Education Researchers*, pp. 52–68. Paris. http://collab.northumbria.ac.uk/2011paris/?page_id=2 CUMULUS.

Roberts, P.H. (2000) Aspects of research concerning design education, in E.W.L. Norman and P.H. Roberts (eds), *Design and Technology Educational Research and Curriculum Development: The Emerging International Research Agenda.* Loughborough: Department of Design and Technology, Loughborough University. http://magpie.lboro.ac.uk/dspace/bitstream/2134/953/1/norman-roberts2001.pdf.

Trimingham, R.L. (2007) *An Exploration of the Roles Values Play in Design Decision-making*, Ph.D. Thesis. Loughborough University, Department of Design and Technology. https://dspace.lboro.ac.uk/dspace-jspui/handle/2134/7975.

Chapter 2

Design Education as the Passport to Practice

MICHAEL TOVEY

Design is what designers do. It is an activity which designers engage in to produce designs. Within design there is a greater emphasis on being able to do it than on designers being a repository of specialist knowledge. This is acknowledged in design education where there is a dominance of design practice in which students engage in the process of tackling design exercises which mimic professional design practice. In our universities and colleges there is a long tradition of teaching design in this way. Students who wish to become proficient as designers devote their time to engaging with design project activity. This develops in intensity and detail and as students become more experienced they are able to tackle progressively more complex design problems. Typically the end goal is that of achieving a level of capability to function as designers in the professional world. That is, they wish to become part of the community of design practitioners. Today it is vital that their education helps them construct a 'passport' to enter this community.

For many design students the portfolio is the physical manifestation of their passport to design practice. With this assemblage of work they demonstrate that they can tackle design problems to a standard which is recognisably that of their professional community. Typically the work in the portfolio is primarily visual, and good representational work including drawing and modelling skills are very important. In this they show that they can think in a 'designerly' way, engaging in a 'solutioning' process. However, it can be argued that demonstrating the ability to engage in creative synthesis is the most important ingredient in the mix which is required to achieve this passport to enter the community of practice.

Communities of Practice

The world of design practice is large with many separate groups designers specialising in different areas. Each of them could be considered as a separate professional group and could constitute a community of practice. Such a community is typically a group of professionally qualified people in the same discipline all of whom negotiate with and participate in a mutually understood discourse. This discourse is both explicit and, very often, tacit and the signs of membership are usually unmistakable.

It is possible to understand this shared discourse in terms of the Community of Practice Theory which was devised by Jean Lave and Etienne Wenger (Lave and Wenger, 1991). It has provided an innovative foundation for many researchers as a social theory of learning which highlights the value of our 'lived experience of participation in the world' (Wenger 2007). Learning takes place through a deepening process of participation in such a community of practice with identities formed from participation.

Wenger characterises a community of practice as displaying a number of key elements. For example, it will consist of a group of people who share a concern for something they do and who interact regularly to learn how to do it better. Such a group is not merely a community of interest, such as people who are interested in certain kinds of films or theatre. Rather members of a community of practice are practitioners, who develop a shared repertoire of resources, experiences, stories, tools, ways of addressing recurring problems in what amounts to shared practice. In pursuing their interest in their domain, members will engage in joint activities and discussions and help each other, sharing information to build relationships that enable them to learn from each other.

Learning within a community of practice is an experience of identity formation. It is not just an accumulation of skills and information, but also a process of becoming – in this case a certain kind of creative and critically minded design practitioner. It is through this 'transformative practice', as Wenger calls it, within a professional community of creative design practitioners that learning can become a source of motivation, meaningfulness and personal and social energy. Wenger helpfully describes features of a well-formed community of practice.

Indicators that a community of practice has formed would include:

1.   sustained mutual relationships – harmonious or conflictual

2.   shared ways of engaging in doing things together

3.   the rapid flow of information and propagation of innovation

4.   absence of introductory preambles, as if conversations and interactions were merely the continuation of an ongoing process

5.   very quick set-up of a problem to be discussed

6.   substantial overlap in participants' descriptions of who belongs

7.   knowing what others know, what they can do and how they can contribute to an enterprise

8.   mutually defining identities

9.   the ability to assess the appropriateness of actions and products

10. specific tools, representations and other artefacts

11. local lore, shared stories, inside jokes, knowing laughter

12. jargon and shortcuts to communication

13. certain styles recognised as displaying membership

14. a shared discourse reflecting a certain perspective on the world.

The emphasis on practice is source of coherence in the community. It is this which distinguishes such communities from, for example, a community of interest, including as it does both explicit and tacit knowledge and information. Frequently this is not overtly stated, and can include a range of elements such as implicit relations, tacit conventions, subtle clues, untold rules of thumb, recognisable intuitions, specific perceptions, well-tuned sensitivities, embodied understandings, underlying assumptions and shared worldviews.

It is through their shared experience of practice that members of such a community develop, share and negotiate their identities. For newcomers the process of assimilation is initially through peripheral engagement. This is progressively legitimised by established members who induct newcomers in much the same way that they were in their turn inducted.

Relations between members are constantly evolving, and each new generation can cause disagreements with more established members as they introduce new ideas and different ways of looking at processes and practices. Communities are not havens of peace and indeed such friction is essential to maintain their energy and vibrancy.

As a community develops its history gives rise to a number of important distinguishing characteristics, such as mutual engagement, joint enterprise and a shared repertoire. There is negotiation between members to share the meaning of the activities which their practice involves which can lead to shared repertoires. These can be in the form of routines, work, tools and methods, or even anecdotes, gestures, symbols and actions. Involvement in a community of practice can go beyond the concern with the professional activities, to include social interaction. There can be an emphasis on not only staying in contact with professional developments within the discipline, but also the latest gossip and rumour. The joint enterprise is based on ownership, with economies of communication and shared understanding leading to common responses and shared practices.

Design Communities

There are many types of designer. We can include architects, industrial designers, design engineers, graphic designers, fashion designers, interior designers, craft designers, furniture designers, jewellery designers and many more. Each of them represents a significant group of professional practitioners and each one could be regarded as a community of practice. Some of the categories are sufficiently large that they subdivide into groups of more specialist designers. Thus for example graphic designers might distinguish between those concentrating on corporate identity, media graphics, or information design. Similarly industrial design contains the large subcategories of product design, and automotive design, and smaller groups such as boat designers.

These professional groupings can be seen in the context refinement of a family of design activities each with its own history and traditions. Walker and colleagues (1989) have developed a representation of the range of design specialisms which gives some sense of their historical development, and shows diagrammatically the interrelationships between design disciplines. It has its roots in traditional craft skills and methods such as drawing, modelling and simulation, to show how it has spread into more specialised activities. It ranges from graphics and fashion, which rely on artistic sensibilities, to science-dependent activities such as engineering and electronics. Some designers may spread across more than one area, and others may be more narrowly active. This helps us to understand the diversity of design and understand its interrelationships and development (Cooper and Press, 1995). I have developed a version of this diagram for use in this context (Figure 2.1).

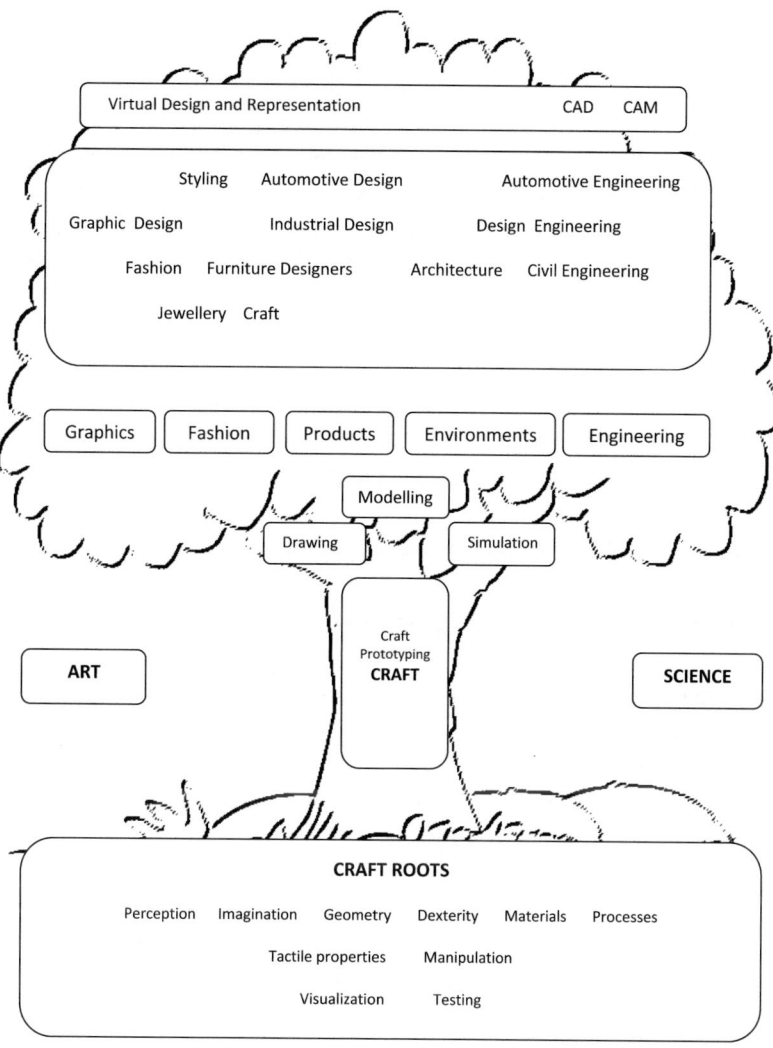

Figure 2.1   The roots of the design community

For key groups there are formal national bodies to which entry is by examination. Thus, for example, for architects there is the Royal Institute of British Architects in the UK, and the Society of American Architects and the American Institute of Architects in the USA. Each country has its own national equivalent for architecture. For a wide range of design professions in the UK there is the Chartered Society of Designers and in the USA there is the Industrial Designers Society of America. Some designers find the Institute of Engineering Designers more appropriate. Most such societies are national and tend to have national membership. The less formal groupings can be international in scope. A powerful example is that of the community of practice of automotive designers.

University life provides such a local community where progress is supported by the desire to become accepted within the greater community of creative design practitioners. It is through this 'transformative practice' (Wenger, 2007), that a professional community of creative design practitioners evolves. Just being in a local community aids acceptance into the larger arena, learning can become a source of motivation and meaningfulness for personal or social energy – in effect an experience of identity formation (Tovey and Owen, 2006; Tovey and Bull, 2010).

This chapter takes as an example the community of practice for automotive designers There are parallel communities within art and design such as fashion and textile design, which echo the main features.

These communities are populated by people who have:

- lifelong ambitions that develop in early childhood

- an evolved and distinct visual language employing symbolic elements at the concept stage

- access to an extended network of professional practitioners who support learning and understanding

- a wish to constantly monitor and attend degree show exhibitions

- capability to offer 'live' projects for student participation.

Figure 2.2 shows diagrammatically the interrelated nature of such a community where focus is finely attuned.

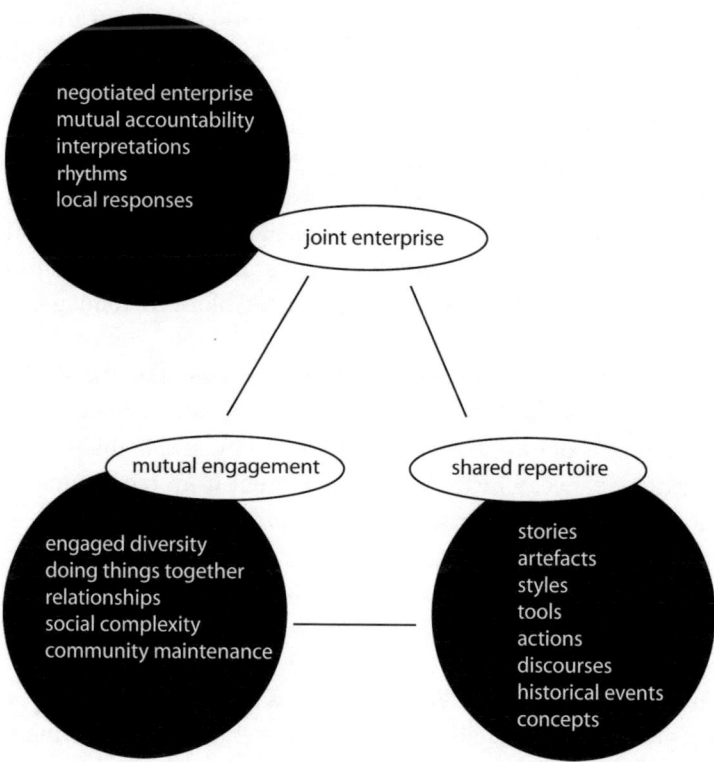

Figure 2.2   The interrelated nature of a community of practice

The Community of Practice of Automotive Designers

The international group of professional car designers is a good example of a community of practice. Such communities are typically characterised by a developing history (Osmond, 2012) and key to the development of automotive design was the contribution which Harley Earl made as Head of Design at General Motors (Bailey, 1990).

It began in the 1920s when Earl, at that time an independent consultant and car body customiser, was appointed to develop a one-off car for the company. His assignment, according to Bailey, was 'to design a car that was to be a sort of people's Cadillac, which would fill the gap between the most expensive Buick and the Olympian Caddy'. Up until Earl's appointment, the American car market had been dominated by Henry Ford's Model T, which was available as a standard generic model for everyone. General Motors noticed customers' increasing habit of trading in their old car for a new model and realised that tapping into customer appeal could give them an edge in their competition with Ford. Earl's background was in designing custom-built cars for the stars of Hollywood. He was seen as someone who could utilise his flair and design expertise to add visual appeal to the company's product. He was given the opportunity to develop a new car to entice customers to their brand by recognising the more sophisticated desires of a second generation of car owners. Completed in 1927, his first production model was the La Salle. It was a major success and on the strength of it General Motors hired Earl full-time and gave him a new department, entitled 'Art & Colour' to manage.

The striking appearance of the company's vehicles from the 1930s until the end of the 1950s was created by Earl and the people working for him. This included times of economic expansion for the USA, and some periods of self-confident extravagance, which are vastly different from the economic realities facing the car companies and their designers today.

One lasting legacy of Earl's approach is the structure of the automotive design process, including the activities which designers engage with in designing new vehicles and presenting their proposals. Much of this has evolved from the systems that Earl set up, and are still in use today, although the representation and modelling systems have changed considerably (Tovey, 2012).

But this is not his only legacy. His contribution was crucial to General Motors' becoming the biggest car manufacturer in the world for much of the twentieth century. He managed not only five separate car design studios, one each for Chevrolet, Pontiac, Buick, Oldsmobile and Cadillac, but also twelve specialist studios. During his period at GM his designs accounted for more than 50 million vehicles. This scale of operation involved the company's employing a significant number of designers, clay modellers, engineers and managers. Although many of these were trained in-house, many also joined from elsewhere, and they stayed at GM for different periods. Many left to take up jobs with other car companies throughout the world with a number becoming the leading designers in these other companies.

As a quite natural process this collection of designers came to function as a community of practice. Many of them had a worked as colleagues with a shared experience of design studios, working methods, design philosophy and project management. Consequently they had a common vocabulary, shared enthusiasms, and often similar attitudes. Naturally this community started out as a predominantly USA-based phenomenon. However, as General Motors was itself increasingly developing as an international and multinational company, its designers were gaining wider experience and becoming international too. And in so doing they communicated with and shared experiences with designers from other companies, and in other countries.

General Motors is less dominant than it was, following the rise of other companies in Japan, Europe and now China. Communications across the globe are instant, and travel is much easier. The community of practice of car designers is international and multi-company and there are car design studios in all of the major industrial countries of the world, and in most of the world's continents. Typically the designers who work in these studios share their passion for automobiles and design novelty. Each time a new vehicle concept is revealed by one studio it causes interest and excitement in the other studios.

Designers are able to stay in touch with developments through the many Internet resources and publications about automobiles. Although during the development of a new design there is usually great secrecy in the company concerned, a great deal of information is shared throughout the industry, and companies often move in similar directions, responding to common pressures from the market and governments.

Each company will have one or more design studios, with its own team of designers, engineers and other specialists. Naturally these designers will often pursue individual career paths which take them from one company to another. This body of people could be said to constitute a community of professional practice. Further it is a fairly visible and high-profile community, and it is one which many students of transport design aspire to join.

For an international community to function it is important that there is communication between its members. Designers are able to stay in touch with developments through the many Internet resources and publications about automobiles. This is supplemented by online resources such as the Car Design News website. This was created by three car designers from the USA and the UK. It contains news from a designer's perspective of developments in car design, with in-depth reviews, an extensive online gallery from all of the major car shows and student exhibitions and competitions, discussion forums, resources and job listings, a large online collection of car designer portfolios, a paid-for members editorial and a car design taxonomy. With over a million hits a year it is a highly effective device for facilitating the community of practice (Osmond, 2012).

Engaging the Profession in Design Education

If the notion of design education as providing a passport to enter the community of professionals is to be acted on then it is essential that design professionals be engaged in the process. At Coventry University the transport and product design scheme is achieved through a number of arrangements (Tovey and Owen 2006).

- A high proportion of the staff has come directly from professional practice. Currently about 65 per cent have come from an industrial or consultancy background. The majority have come from the transport industry.

- There is industry involvement in the delivery of the programme. In a four-year scheme, typically there is industry-related project work from year two onwards.

- Students have the opportunity to experience professional placements and internships. This happens usually in year three and often involves work outside the UK.

- The department engages in client-funded design work and applied research. Highly relevant examples are the Benero car launched at the Geneva Motorshow, for which the department undertook the concept design and styling work, and the design-based Microcab hydrogen fuel cell vehicle, which is a spin-out company located on Campus.

A key and distinctive strength for this method of fostering close engagement with leading industrial designers from the transport industry is that it provides a useful model for emulating professional practice through live design projects. This policy is appropriate to the work required within the discipline community and the effective use of industry engagement in its delivery constitutes a significant access point for students to the community of designers.

Student study is supported with extensive opportunity for engaging with real-word design problems in each course. Industrial collaborations feature particularly in the second half of year two, throughout the 'Professional Enhancement' year (the third year of our four-year undergraduate course) and the postgraduate diploma stage of the Masters provision. Industry experiences and contact with professionals in the field tend to build incrementally until near professional engagement is displayed, a time where student and designer interact equally.

Distinguished representatives of the design industry are invited each week to give a lecture, open to all staff and students, keeping open the two-way interchange of ideas necessary for developing a shared view of design direction. Eminent design professionals are also engaged to make input to courses at a theoretical and practical level in collaborative projects. Industry provides design briefs for students that normally include specific requirements, for example, the need to:

- incorporate a particular vehicle platform or donor parts

- use a particular process and/or material

- meet the functional needs of a particular subset of the population

- propose an innovative vehicle concept.

Such constraints are a spur to creative thinking – students are required to review the task critically, and to question the brief as means of exploring the creative possibilities contained within it. Thus, formulation of the project brief becomes a dialogue and then contract between client and designer.

Tovey, Porter and Newman (2002) discuss how automotive stylists are expected to display visual flair within a controlled and yet changing formal vocabulary, the group culture is such that they recognise a shared but exclusive language. Their ability resides in tacit knowing – an apparently subliminal appreciation of the shapes acceptable for a car design and trends in automotive styling domains central to their work. Studio design representations, CAD and physical models clearly show the design culture and its dependence on visual language. Because such groups are small, and they may work together for a number of years, their group language may be idiosyncratic and atypical – a prime indicator of a mature community of practice.

The strongly held view is that spatial understanding, creativity and innovation are central to the design process, while the engineering and the ergonomic credibility of products are crucial to success. The proposed design must be manufacturable as well as meet the needs and desires of the people for whom it is intended. The close involvement with industry ensures the continued real-world relevance of the design programmes.

Thus there are effective mechanisms for ensuring the professional relevance and industry orientation of the courses of study. This has ensured that within the programme work is presented to a high level of professional polish. Typically students leave their studies with portfolios of work whose presentation standard matches well the work of the industry design studios. That aspect of the portfolio's function in providing a passport to practice is being met quite proficiently.

Feedback from our industry colleagues is that the element most needed in such portfolios, the creative sparkle, is also the most difficult to achieve. Clearly professional competence in presentation is not sufficient on its own. There needs to be a complementary engagement with developing creative design-thinking capability.

Conclusions

Central to the notion of a 'passport to design practice' is the recognition of the existence of groups as communities of design practitioners. Where such communities are national and wear the badge of a professional body or society they are easy to identify and quite visible. However, there are large less formal international communities of design practice whose influence can be just as profound. It is proposed that for each specialist group of practicing designers there is either explicitly or implicitly a community of professional practice.

Each of these communities will have its own specialist history, technology, skills and expertise. Gaining entry to such a group will involve demonstrating familiarity with these domain-specific skills and capabilities and being on top of specialist information. But each of them also requires capabilities which are generic across design specialisms.

For those embarking on practice-based design education where they will spend much time tackling design problems, these shared capabilities are part of the core competency which they need to master. The designing involves a solution-focused creativity which is evaluated through access to analytical processes. It is essentially the deploying of visuospatial intelligence in solving design problems. This visual thinking is externalised as design drawings and representations, both as a key ingredient in the design process, and when bundled together in a portfolio, to become the entry ticket to the community of design practice. In other words it is their passport to enter the community of practice.

Expertise in design thinking is crucial to all sorts of design activity, but because it is generic it also has wide application outside design. It is a highly transferable skill meaning designers can be very effective members of teams tackling a range of different problems.

References

Bailey, S. (1990) *Harley Earl*. London: Trefoil Publications Ltd.

Cooper, R. and Press, M. (1995) *The Design Agenda: A Guide to Successful Design Management*. Chichester: John Wiley and Sons.

Lave, J. and Wenger, E. (1991) *Situated Learning Legitimate Peripheral Participation*. Cambridge: Cambridge University Press.

Osmond, J. (2012) Passports to a community of practice, in M. Tovey, *Design for Transport: A User-centred Approach to Vehicle Design and Travel*, pp. 335–352. Design for Social Responsibility Series. Farnham: Gower.

Tovey, M. (2012) *Design for Transport: A User-centred Approach to Vehicle Design and Travel*, Design for Social Responsibility Series. Farnham: Gower.

Tovey, M. and Bull, K. (2010) Design education as a passport to professional practice. Presentation at EPDE 2010, September, Trondheim, Norway.

Tovey, M. and Owen, J.K. (2006) Entering the community of practice of automotive design, in I. Horvath and J. Duhovnik (eds), *Proceedings of TMCE 2006 Tools and Methods of Competitive Engineering International Symposium*, pp. 715–724. TMCE 2006, in Ljubljana, Slovenia. Delft: Delft University Press.

Tovey, M., Porter, C.S. and Newman, R. (2003) Sketching, concept development and automotive design, *Design Studies*, 24(2002), 135–153.

Walker, D. et al. (1989) *Managing Design: Overview Issues, P791*. Milton Keynes: Open University Press.

Wenger, E. (2007) *Communities of Practice Learning, Meaning and Identity*. Cambridge: Cambridge University Press.

Chapter 3

Designerly Thinking and Creativity

MICHAEL TOVEY

What designers do in a variety of contexts is create designs and solve design problems. It is in this most obvious sense they are different from other professionals. One contention in this chapter is that in order to function effectively as designers they must engage in creative designerly practice. This is a core capability which is shared across different types of designer, and based upon creative design thinking. The process which designers go through is at its simplest level generic. There is a movement from an initial brief through a combination of processes at the end of which there is a credible new design proposition.

Although much of what designers need to know is specific their specialist area, making it distinctive and giving it a specific identity, there are core similarities across specialist design areas. After all they would not be called 'design' if they did not share with each other a common preoccupation with solving design problems. A good generic definition of design which signals what this core activity is appeared in the Cox Review (Cox, 2005).

The Cox Review of Creativity in Business appeared in 2005. In it Sir George Cox defined design as follows:

> *Design is what links creativity and innovation. It shapes ideas to become practical and attractive propositions for users or customers. Design may be described as creativity deployed to a specific end. (Cox, 2005: 2)*

Designerly Knowing

Designing is a complex and sophisticated activity, and one which is usually regarded as inherently creative. It remains one of our least well understood cognitive powers and one of the most difficult to teach (Lawson and Dorst, 2009). It is both creative and unpredictable. Research in design has shown that there are many ways of designing well and successfully: indeed current thinking about design sees every single project as unique and special.

One of the important and distinguishing characteristics of design thinking was identified by Darke (1979) as the primary generator. In her study of how architects work she noticed that they produced a major design idea very early in the process, long before they had fully understood what were very complex problems. Rather than engaging in a detailed preliminary analysis of the design problem designers tended to use solution-based approaches. She called this early production of a design idea the 'primary generator' which was to drive a process of conjecture and analysis. This approach to designing, in which there is a production of a solution proposal at a very early stage, has been characterised by Lawson as the solution-led approach (Lawson, 2005).

This core process makes the design activity different from the processes of social or scientific analysis which are central to other disciplines. It implies a whole way of understanding the world and responding to it. This has been characterised by Cross as the 'Designerly Way of Knowing'. (Cross, 2006), Cross identifies this generic design capability as containing five aspects:

- Designers tackle ill-defined problems

- Their mode of problem-solving is solution-focused

- Their mode of thinking is constructive

- They use codes that translate abstract requirements into concrete objects

- They use these codes to both read and write in the object languages.

It is in the character of design problems that they tend to be ill-defined, ill-structured, or 'wicked' (Buchanan, 1992) and designers may not have all the information necessary to solve them. To cope with this lack of information, experience indicates that the quick production of a draft solution will allow a definition of the limits of the problem and the provision of a basis for developing an idea or ideas further. To quote Cross:

> *In order to cope with ill-defined problems, the designer has to learn to have the self confidence to define, redefine and change the problem-as-given in the light of the problem that emerges from his mind and hand. People who seek the certainty of externally structured, well defined problems will never appreciate the delight of being a designer. (Cross, 2006: 7)*

Cross makes a number of useful and relevant observations about the design cognition process, noting that designers are solution-focused not problem-focused. The designer's attention oscillates between the problem and its solution, in an appositional search for a matching problem–solution pair, rather than a propositional argument from problem to solution. It is probable that design thinking operates with two simultaneous interacting cognitive styles being employed. Thus it would be expected for example that an analytic, linear strategy would be at work in the process of data generation and organisation, to yield a design specification, and also in the evaluation of design proposals. In parallel with this a synthetic–holistic strategy, used in the generation of solution conjectures, would be the integration of visual relationships and the physical representation of the design as drawings or 3D models.

In the solution-led approach the production of a solution conjecture at an early stage in the process could be said to facilitate the re-examination of the problem by providing the spectacles through which to look at it. The designer is able to tell where they need more data because without it the design cannot move forward. In some areas of design this solution-focused strategy is fully formalised in the way in which the design activity is managed, for example at an early stage in the process there will be a requirement for a 'Concept Design' which is the designers' attempt to provide a sketchy representation of what the finished design might be, or might look like. If the designer or design manager sees the concept as providing a basis for proceeding then the structure of the rest of the process falls into place. This is the solution-led approach, which has, at its core, the process of moving from an abstract statement to a visual object. The designer learns to think in a sketch-like form, in which the abstract patterns of user requirements are turned into the concrete patterns of an actual object. Thus the designer uses a code to effect this translation from individual, organisational and social needs to physical artefacts. This is the use of the language of designing, employing its translation codes, and is the match of the analytical statement to the holistic solution. The manifestation of this outcome will be a visual representation, a drawing, a 3D or virtual model.

Types of Intelligence

All of the types of designer we are considering concern themselves with the design of objects, either in two dimensions or three. They all work in the real world engaging with the creation of visuospatial entities – that is, physical objects or virtual objects perceived as physical objects.

Howard Gardner's theory of multiple intelligences (Gardner, 1993) fits well with a number of aspects of design thinking and it has provided a useful focus for design educators. Gardner identified seven distinct types of intelligence: linguistic, logical–mathematical, musical, bodily–kinaesthetic, spatial, interpersonal and intrapersonal intelligence. Each of these is distinct, meeting the seven criteria he specifies. Interestingly these include the association of an area of the brain with each of the specialised intelligences.

Spatial intelligence or spatial understanding is the form of intelligence most obviously linked to design thinking. However, it is a wide-ranging capability which includes spatial orientation, and the ability to analyse the spatial problems which psychologists use. Gardner's work draws attention to 'more abstract and elusive' spatial capacities, such as sensitivity to the various lines of force that enter into a visual or spatial display and the identification of resemblances that may exist across seemingly disparate forms. The metaphoric ability to discern similarities across diverse domains also derives from spatial intelligence and is an important attribute of design thinking.

Gardner uses terms such as the 'language of space' and 'thinking in the spatial medium' which have obvious relevance to design thinking. Design activities such as drawing and modelling are directed to a more sophisticated visuospatial understandings of the complexity of surfacing and three-dimensional form. They are key ingredients in developing visuospatial capability, representing the object language in a designerly way of knowing.

Design Thinking as Dual Processing

There has been much research into how the brain functions. Some of this has focused on the different characteristics of the two halves of the brain, or cerebral laterality. Many researchers in this field have characterised the two hemispheres of the brain as separate information processors and encoders. There is strong evidence that underlying the left hemisphere's dominance for expressive speech and the right hemisphere's dominance for manipulospatial activities are different processing modes. Typically the modes are characterized as analytic–synthetic, linear–holistic, serial–parallel or focal–diffuse for the left and right halves of the brain, respectively. This dichotomy is attractive because it seems to correspond with the different types of cognitive style identified by psychologists in problem-solving procedures. It also seems to correspond with the analysis–synthesis dichotomy which has been identified as the basis of the design process.

It is clear that for anything other than very simple mental operations, both halves of the brain are involved, as has been shown in magnetic resonance imaging (MRI) maps of cerebral activity during experimental tasks. It would seem that the two processing modes are typically employed at the same time and interactively, and that a more complete understanding of any particular problem arises from the matching of initially separate simultaneous mental operations.

These two interacting lateralised mental operations can be used to map out design thinking and help understand it (Figure 3.1). I have called this the dual processing model of the design process (Tovey, 1984). Within this there is the assumption that the two halves of the brain will both be involved in solving the design problem, each half working in its own preferred information-processing mode, each tending towards its favoured modelling language, the left in words and symbols, the right in drawings, 3D constructions and digital CAD models. The design process will be concluded successfully when the two processors are in agreement over a solution.

The essence of this model is the interaction of the two modes of thought, each stimulating and modifying the other, both crucially involved in the evolution and resolution of a fully detailed design proposal. Although such a model is of course speculative, it does provide a framework within which different design approaches can be accommodated. The relative emphasis given to serial–analytical and to simultaneous holistic thinking varies both between designers and between types of design problem. For example, engineering designers may give first priority to analysis and the derivation of a specification, whereas product designers may concentrate more on the holistic processes used to derive a design concept presented as a drawing or a 3D model. Nonetheless it is assumed that the design process will always involve both modes of thinking, and that it is their relative proportions which will vary.

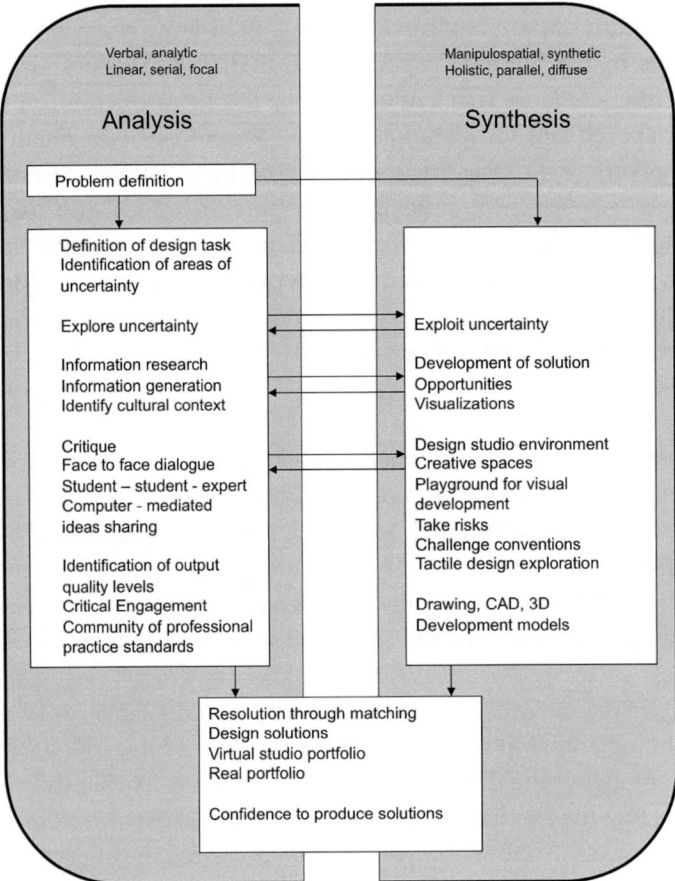

Figure 3.1   Dual-processing model of the design process

In any analysis of thinking there are many dichotomies which seem to parallel the dual processing model. These include the distinctions between convergent and divergent thinking, and that between reflective and impulsive thinking. One of the most fundamental is that between verbal and non-verbal thought, and as noted above, design thinking contains a high proportion of non-verbal thinking.

Those who have contributed most to productive techniques in thinking seem to have had a need to describe it in terms of such dichotomies, and one of the most useful has been Edward De Bono's split between vertical thinking and lateral thinking (De Bono, 1970). This has much in common with the left-hemisphere/right-hemisphere split of the dual processing model and also relates to other aspects of design thinking. Lateral thinking is a term invented by De Bono and he contrasts this with traditional, logical thinking which he calls vertical thinking.

Vertical thinking is very useful, particularly when analysing a problem, or putting forward an argument. Typically it involves making yes/no decisions at each stage, selecting and discarding material, being right at each stage. Although it is powerful and useful, it can also be limiting. It is the equivalent of digging a hole by making the hole you already have deeper. Lateral thinking is concerned with digging as many new holes as possible, for the solution may not be in the direction in which you are already digging. There are many techniques for facilitating lateral thinking. Some are concerned simply with overcoming the limiting effects of vertical thinking, by challenging assumptions or by suspended judgement. Others are more provocative, creating new combinations, concept changes and idea reversals to encourage innovation.

The Analysis Synthesis Dual-processing Model

The essence of the dual-processing model (Tovey, 1984) is the interaction of the two modes of thought, each stimulating and modifying the other. As the designer's attention oscillates between the problem and its solution, in an appositional search for a matching problem–solution pair, the left hemisphere preference is for local, linear, narrowly focused attention, the right hemisphere's preference is for simultaneous, broad, global and flexible attention (McGilchrist, 2009).

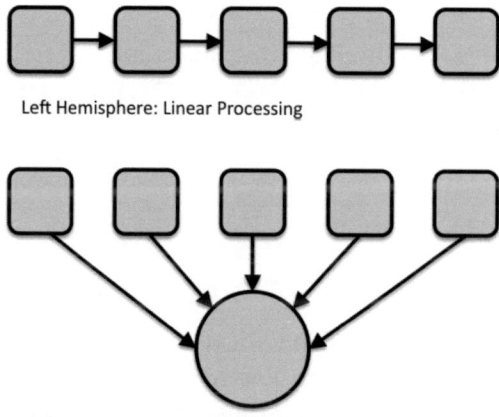

Left Hemisphere: Linear Processing

Right Hemisphere: Simultaneous Processing

Figure 3.2   **Linear and simultaneous processing**

A recent and very fruitful development of De Bono's ideas has emerged from the Institute of Practitioners in Advertising (IPA). The idea for 'Diagonal Thinking' was initiated by Hamish Pringle (2009) who has a senior role in the IPA. The concept underpins a tool designed to aid the recruitment of talented people into the creative industries. A self-assessment psychometric test is offered as a way of signalling whether the applicants are able to employ the appropriate combination of capabilities.

The core notion of diagonal thinking is that to be effective in creative professions a combination of both linear and lateral thinking is required. The picture used of diagonal thinking is as a diagonal vector resulting from the matching of vertical and lateral thinking axes (Figure 3.3). This is an attractive and useful way of showing the kind of agreement between linear and simultaneous thinking proposed in the dual-processing model. It would seem to play a fundamental part in design thinking.

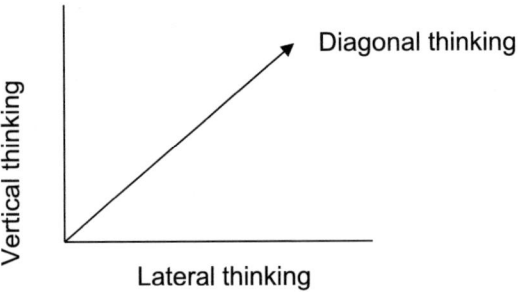

Figure 3.3   The concept of diagonal thinking

A key question is how this balancing, matching and resolution of thinking modes can be achieved. This would seem to be the necessary skill which is required for design creativity.

Parallel Lines of Thought

Lawson and Dorst (2009) talk about designers employing parallel lines of thought. Designers need to be able to develop parallel lines of thought about a problem–solution situation. Each line of thought develops to respond to a frame in order to restrict the view of the problem and to rely on a primary generator to develop ideas about the solution. It seems probable that highly creative designers may be able to sustain several of these parallel lines of thought and allow them to be incompatible or even apparently irreconcilable for extended periods in a design project. Indeed it may even be the case that creative reframing of a situation allows for a new view in which the various lines of thought can be incorporated in a higher-level set of ideas. The ability to think along parallel lines, deliberately maintain a sense of ambiguity and uncertainty and not to get to concerned with a single answer too quickly seem to be essential design skills.

In the process of creating a design there seems to be a need for these two or more simultaneous tracks of thought covering both logical analysis and creative thinking. For a number of types of designing the reframing of the context through analytical activities such as data acquisition, problem definition and solution evaluation need to be tackled in an explicit and objective manner. The design brief and specification can be the initiating force which both precedes the development of design proposals and functions as a checklist to evaluate them.

However, if the design is to be creative then the re-framing needs to include emphasis on the identification within the design problem of whatever will allow the early generation of a solution conjecture which can take the form of producing a concept design. Lawson and Dorst note that conceptual design can be described as the art of seeing the design situation in multiple ways or 'seeing as'. Designers are used to performing this little dance around the problem, taking stabs at it from different sides. Once more design can be seen as a mixture of creativity and analysis. It is not one way of thinking but several, mixing rational, analytical thinking and creativity. This inherent schizophrenia is a defining characteristic of design and directly leads to the peculiar way of working that is a common trait of practice throughout the design professions.

For the individual designer or design group this can expose a real dilemma often keenly felt in design practice. Confronted with a design problem one might tackle it in either a problem-focused (analytical) or a solution-focused (creative) way. This can be a difficult choice for a designer as the project can fail if the balance is wrong. Being too analytical can lead to an unnecessary limitation of the solution space, while being too creative and generative can launch a journey into nothingness.

This use of a blend of different thinking styles makes it difficult for many people to understand design. But to designers, these thinking styles are so intimately connected in a design project that they seem almost merged into one way of thinking. When steeped deeply in your design activity you just keep switching between analysis and creativity, between 'problem' and 'solution' without any effort. As Lawson and Dorst note, 'in practice it is often devilishly hard to distinguish between them' (2009).

The key to effective design thinking is to both allow these parallel activities and develop productively, and to develop techniques for mediating between them. Reflective practice could be the most effective way of doing it.

Reflective Practice

Donald Schon (Schon 1991) developed the idea of reflective practice. He applied it to professions such as design and contrasted it with technical rationality in incorporating a developed level of tacit knowledge. It involves being knowledgeable in a special way, so that in our actions we demonstrate an internalised understanding. Skilful action often reveals a 'knowing more than we can say'. Thus the reflection needs to develop into 'reflecting in action'.

Schon describes design (and work in the other professions he studied) as a process of 'framing' the problem (a form of 'seeing as'), performing 'moves' towards a solution, and the 'evaluation' of these moves, that might lead to new moves or to seeking a new frame. He describes framing as involving selectively viewing the design situation in a particular way, in order to handle complexity and contradictions.

Designing is seen as a conversation with the materials of a situation. The process moves from possibilities to firm conclusions which may require the reframing of the approach as a radical strategy. The dialogue consists of a proposition–evaluation process in the cyclic evolution of both the understanding of the brief and the development of a solution proposal. There is an obvious tension between a problem view and a solution view of the situation. This is at the heart of the way designers think. And creative designers may tend be more solution-focused in a process which is challenging and frustrating, but also satisfying and compulsive.

Schon observes that designers have the task of negotiating reconciliation between these two views of the situation they are dealing with. The problem view on the one hand is expressed in terms of needs, desires, wishes and requirements. The solution view on the other hand is expressed in terms of materials, forms, systems and components. Since these two views share no common language this reconciliation requires some very clever mental tricks indeed. In this view of the design process we do not see designing as a directional activity that moves from the problem through some theoretical procedure to solution. Rather we see it as a dialogue, a conversation, a negotiation between what is desired and what can be realised.

Schon's particular contribution is to identify reflection as the process by which this reconciliation can be brought about. Designers all have their own unique background and collection of skills, attitudes, values and interests and each design problem is unique, so designers will be working on their own set of problems and circumstances. But Schon also notes that it is generally accepted that being able to draw is likely to help a designer to perform better. The designer has a conversation with the drawing in a three-stage process of framing, moving and evaluating.

Developing Design Thinking

In the picture of designing which emerges here it is seen as a process of developing two or more interacting parallel lines of thought, including one concerned with analysis and evaluation, and at least one other concerned with constructive solution proposition. The route to a resolved design involves the matching and harmonising of the parallel lines of thought. At a simple level if the analytical evaluative side of the process includes a requirement for novelty and innovation, then it will serve to stimulate the constructive side to develop original design ideas. In that sense both processes are crucial to designing as a creative activity. It is suggested that reflective practice offers an approach through a conversation between these two views, as a re-framing process which allows tacit understanding a role in matching the modes of thinking.

Although this may sound somewhat mysterious and esoteric as a process, the long experience of design educators is that there are many ways of inculcating this design ability. Indeed it is possible to unpack it and regard it as a skill, many aspects of which can be taught.

Edward de Bono asserts that to regard thinking as a skill rather than a gift is the first step to improving that skill. On the whole, it must be more important to be skilful in thinking than stuffed with facts (De Bono, 1970).

Teaching Design

In order to teach design effectively it would seem to be necessary to incorporate approaches which accommodate the development of these parallel lines of thought, with space for the reflective techniques which will allow for their harmonisation in the resolution of the design. This could be regarded as the key to producing design students adept at this complex but engaging skill. Lawson (2005) proposes an approach to doing this and identifies the key components in designing as:

- Formulating

- Representing

- Moving

- Bringing problems and solutions together

- Evaluating

- Reflecting.

Formulating is about ways of understanding design problems. It includes identifying, re-formulating and giving structure to ill-structured or wicked problems. Through the framing process it is suggested that the problem can be viewed selectively in a particular way for a period, before moving on. Such a framing process in which they may be viewed in a particular way for a period, acts as a kind of window on the problem space.

Representing is about externalising design ideas, typically as drawings or models of one sort or another. Designers interact with such representations in a conversational way, and sketching may be regarded as a skilled form of visual conversation. For some design projects multiple representations may be required, each of them offered to the design problem as both a way of understanding it and as a way of developing the design.

Moving is about creating solution ideas, generating solution concepts, using the primary generator as the acknowledged label for the early production of a solution proposal based on partial information. Such new ideas may be developmental and interpretive, or radical.

Bringing problems and solutions together: this acknowledges that the problem and solution are inseparable, and that there is no clear order of appearance. Thus briefing is a continuous process and at the core of the design process are parallel lines of thought.

Evaluating involves the deploying objective and subjective evaluations, both of the solution proposals and of elements in the problem space. However, one of the most important skills for the designer is knowing when to suspend judgement so as to allow the ideas to flow.

Reflecting is the key process for the reflective practitioner. It can consist of reflection in action and reflection on action, a process of standing back and reviewing the situation. Through experience the designer will develop guiding principles as well as collecting precedents or references.

These are all core components which would need to be addressed in any situation where design was being taught. However, they are so closely integrated with each other that they cannot be considered as a curriculum of separate topics. What is required is for them to be addressed in integrated design project work in which design skills can be learned, practised and improved.

The arrangements for inculcating design capability in students have conventionally depended on four pedagogic elements:

- the studio

- the design tutorial

- the library

- the crit.

The studio is typically a room in which students engage in design practice. It may be regarded as not just a physical space but also a social and cultural entity. At its best it can be a marketplace for ideas, and as such it is one of the defining characteristics of design education. Often students have individual work spaces, and may have a significant amount of autonomy over how they organise their time and structure activities. It is the arena in which there is the opportunity to achieve the integration of ideas which is at the core of design synthesis. It is also the place where they can mimic professional design activity.

The design tutorial is the individual interaction between the student and tutor. The intention is to progress the design project and thus to develop the student's design capability. Inevitably the tutor may become a kind of collaborator in this process. Conventionally the tutor may take on the role of teacher focusing on the development of the student's ability, or that of consultant, focusing on the development of the design.

The design library is a store of useful information, including for example technical information, or marketing data, necessary to support the design. Unlike the library for a conventional subject, however, it also functions to provide examples of design practice. These may be historical pieces of design work, or they may be the most recent work by the best contemporary designers. The library functions to support both the evaluative strand and the solutioning strand of the student's thinking.

The design crit is sometimes referred to as the 'review' or 'design jury'. The idea is straightforward, students display their design work and a number of staff sit around and criticise this in a public forum with other students listening in. At its best this is a constructive process with encouraging comments which are intended to assist the students in taking forward their work. However, more often it takes place at the end of a project when it becomes an assessment tool.

These are the arenas in which the core design processes of formulation, representation, movement, synthesising problems and solutions, evaluation and reflection need to take place. The tutor's role is to develop the student's ability in these areas using the opportunities of the tutorial and crit. The environment with its two key components of the studio and the design library is key to their success.

Conclusion

The proposition in this chapter is that designing involves a peculiar and particular blend of thinking processes, which are the distinguishing characteristics shared by different sorts of designer. This could be labelled the designerly way of knowing, making use of various forms of intelligence, particularly visuospatial thinking. Its complexity arises in part from its being used to tackle 'wicked problems', those questions which are not precisely formulated and developed. It involves a use of parallel lines of thought which can be corralled into two streams in a dual-processing model which aligns them with the preferred thinking styles of the two halves of the brain. They deploy respectively serial and simultaneous cognition. The consequence of regarding these as equivalent implies an early production of a proposed design in a solution-led approach. This is fundamental to its being a creative activity. Although this is a beguiling and apparently mysterious process much of it can be regarded as a skill which can be taught. Lateral thinking and diagonal thinking offer examples of how such thinking skills can be taught and developed. At the core of the designerly way of knowing is a conversation between these modes of thought. Developing the skill in undertaking this conversation involves reflective practice, in a process which engages tacit knowing and reframing the problem, and the solution. It is possible to accommodate these approaches in a teaching strategy which itself involves several parallel activities. These are devised to allow to the space for reflective design. The key elements in design teaching practice which are necessary for this can be identified as the studio, tutorial, library and crit. Each of these is a traditional component, and using them effectively depends on the approach being informed by a deep understanding of the designerly way of knowing.

References

Buchanan, R. (1992) Wicked problems in design thinking, *Design Issues*, 8(2), 5–21.

Cox, G. (2005) *Cox Review of Creativity in Business: Building on the UK's Strengths*. London: HMSO.

Cross, N. (2006) *Designerly Ways of Knowing*. London: Springer-Verlag.

Darke, J. (1979) The primary generator and the design process, *Design Studies*, 1(1), 36–44.

De Bono, E. (1970) *Lateral Thinking: A Textbook of Creativity*. Ward Lock Educational Press, London.

Gardner, H.E. (1993) *Frames of Mind: The Theory of Multiple Intelligence*, paperback with tenth anniversary edition with new introduction. New York: Basic Books.

Lawson, B. (2005) *How Designers Think*. Oxford: Architectural Press.

Lawson, B. and Dorst, K. (2009) *Design Expertise*. Oxford: Architectural Press.

McGilchrist, I. (2009) *The Master and his Emissary: The Divided Brain and the Making of the Western World*. Yale University Press, Connecticut, USA.

Pringle, H. (2009) *Diagonal Thinking*, IPA website. http://www.ipa.co.uk/Page/ Diagonal-Thinking-Introduction.

Schon, D. (1991) *The Reflective Practioner*. Farnham: Ashgate.

Tovey, M. (1984) Designing with both halves of the brain, in *Design Studies*, 5(4), pp. 219–228. Butterworth and Co. Guildford.

PART 2
KEY DEVELOPMENTS
IN DESIGN PEDAGOGY

Chapter 4

Fostering Motivation in Undergraduate Design Education

STEVE GARNER AND CHRIS EVANS

The role of student motivation in education has long been appreciated and the nature of motivation has been variously deconstructed to reveal, for example, the need to expose learners to challenge, risk and reward. However, many of those responsible for creating effective learning environments still view motivation as one of the black arts of pedagogy. This is particularly true in design education that today presents some distinctive difficulties and opportunities, partly because there have been huge changes in learning and teaching. Fostering motivation today is a very different challenge to even a decade ago.

This chapter, aimed at both learners and teachers, examines the vital stimulus of motivation in undergraduate product design education. It seeks to illuminate how students might develop their motivation through strategies such as stimulating conflict, embracing failure and effective self-management that are sympathetic to design ideation and creative evaluation. In doing so it shines a light on the work of John Keller, who neatly provides a framework for understanding and applying motivation. In essence motivation is key to the creative and analytical thinking that underpins designing.

Introduction

Those leading design education in universities place great emphasis on developing skills and knowledge, and yet many expect students to automatically possess the necessary motivation for operating across today's design practice. Sometimes those who create design education assume their students must have the same drive and enthusiasm as themselves, while others assume that a hunger for success in the form of assignment grades or career opportunities is sufficient motivation. One might imagine that most designers are motivated by money, but the most powerful rewards in design are often those associated with being part of successful innovation, working as part of a team to get a product into the marketplace where it's well received. It is here that undergraduate design courses can overlook such emotional motivation and, even worse, create irrelevant reward systems. Motivation is not some optional add-on. It has a function equal in significance to other intellectual and practical skills and knowledge. Motivation is not a vague, passive force; it can be understood, shaped and developed. In fact, designing demands a constant refreshing and renewal of motivation, and this has implications for design curricula and the sort of blended learning experiences that are created for students.

This chapter makes a case for prioritising the development of motivation in young designers. This is all the more urgent as forces conspire to erode motivation by swamping design tasks with information. Designers need support for the agile navigation of the world of design. We need learning experiences that tap into students' natural motivations but which professionalise motivation to create a resilient, informed and sustainable capacity. Since motivation is not one distinct force but is shaped and coloured by numerous cognitive forces and emotions, it seems logical that any attempt to develop motivation should acknowledge its diversity.

Understanding Motivation

An investigation into strategies for developing motivation will, most likely, bring the reader into contact with a huge back catalogue of research into effective educational practice. It is almost impossible to disentangle sound and effective pedagogy and an awareness of the importance of motivation. We'll avoid a full resume of the roots of understanding human desires, which can be traced back to Plato should the reader be so inclined, and instead join the story in the late twentieth century when motivational research, pedagogy and educational technology collectively created a new vortex of pressures and opportunities. By the 1980s it was popular to itemise short lists of good practice in education. This example by Chickering and Gamson is one example among many referring to undergraduate education. They propose that good undergraduate education:

1.   Encourages contact between students and faculty

2.   Develops reciprocity and cooperation among students

3.   Encourages active learning

4.   Gives prompt feedback

5.   Emphasises time-on-task

6.   Communicates high expectations

7.   Respects diverse talents and ways of learning. (Chickering and Gamson, 1986)

While the university sector was documenting good educational practice that same practice was undergoing seismic change as a result of emerging educational technology which allowed learners and teachers to create radically new learning experiences that, in theory anyway, addressed learners' needs and offered cost-effective solutions to providers through stimulating new interactions. As we entered the 1990s the fit between what was possible and what was desirable was shaped by research from the field of psychology, and particularly educational psychology, on motivation. To a greater or lesser extent all subjects were caught up in this vortex of change but perhaps those areas of the curriculum that were involved in design education faced, and still face, the greatest challenge because of the remit to develop creativity.

The synergies between motivation and creativity stimulated various researchers in the 1990s. (Sternberg and Lubart, 1991; Woodman and Schoenfeldt, 1990) Intrinsic motivation, that is, *motivation that is driven by an interest or enjoyment in the task itself, and exists within the individual rather than relying on external pressures or a desire for reward* was held to be highly influential on creativity but difficult to foster in the classroom. Extrinsic motivation, that is *motivation referring to the performance of an activity in order to attain an outcome, whether or not that activity is also intrinsically motivated* accorded with conventional strategies of educational practice such as setting grade bands, fostering competition and applying punishments, but its value for stimulating creativity was less certain. By the 1990s a two-pronged pedagogical dilemma had emerged: first, the importance of an enjoyable and supportive learning environment was recognised, including supporting the relationship between learner and tutor and between learners, and yet there were so many new ways of configuring learning environments as a consequence of digital technologies. Second, while management and assessment of learning was key it was recognised that learning should allow students some measure of freedom to determine their own goals and the strategies by which they achieve them in a broadly defined task. The dilemma here concerned how to facilitate learners' ability to self-regulate and assess their own outputs and process in ways that were equitable and transparent. This radical shifting of the emphasis onto an active learner emerged simultaneously with a mission to transform clearly defined creative tasks into experiences that were fuzzy and open-ended (Liu et al., 2012).

As suggested above, the understanding of the 1990s was founded on some seminal earlier work. For example John Keller's work identifying the principles of motivation underpinned that of many later researchers. In his 1979 paper we see an early classification of motivational concepts:

> *In order to have motivated students, their curiosity must be aroused and sustained; the instruction must be perceived to be relevant to personal values or instrumental to accomplishing desired goals; they must have the personal conviction that they will be able to succeed; and the consequences of the learning experience must be consistent with the personal incentives of the learner. (Keller, 1979)*

It says something about the robustness of his mapping of motivation that almost three decades later Keller's notion of motivation to learn remained largely the same as these five principles reveal:

Motivation to learn is promoted:

1. when a learner's curiosity is aroused due to a perceived gap in current knowledge

2. when the knowledge to be learned is perceived to be meaningfully related to a learner's goals

3. when learners believe they can succeed in mastering the learning task

4. when learners anticipate and experience satisfying outcomes to a learning task

5. when learners employ volitional (self-regulatory) strategies to protect their intentions (Keller, 2008).

This is also offered as valuable for 'maintaining' motivation.

These five principles provide a relevant foundation for reappraising how motivation is fostered in undergraduate design education today and they form the focus for the next section.

Motivation and Designing

1. MOTIVATION TO LEARN IS PROMOTED WHEN A LEARNER'S CURIOSITY IS AROUSED DUE TO A PERCEIVED GAP IN CURRENT KNOWLEDGE

The key term here is *curiosity* and it is the lifeblood of designing. The creative process is a continual mapping of the known and unknown using various representations to help us bridge the two. It's curiosity that underpins and energises the search for solutions and without curiosity the process breaks down. It's easy to take curiosity for granted – we too easily accept that curiosity is fugitive, intangible and personal. We can easily assume some people have got it while others haven't. But work by researchers such as Keller reveals curiosity as something tangible, something that can be defined and developed. Central to developing curiosity are learning and teaching strategies for gaining attention. Importantly these can involve a deliberate provocation of conflict or other destabilising techniques. They can be operated through verbal critiques in individual or group contexts or they might be integrated into studio practices of making, analysing or imaging. Far from being negative, challenge or conflict can positively influence attention and thus stimulate a sense of inquiry in the learner. It also assists development of a capacity for managed risk-taking. However, no matter how creative a learning environment is, and no matter how much it fosters curiosity or creativity, its effect will diminish over time. The lesson for tutors is to keep it fresh through constant development and change.

At its best, design education has successfully melded a taught curriculum with a flexibility that allows students to search for gaps in current knowledge and encourages them to follow lines of thinking that haven't necessarily been mapped in the teaching. Of courses, this can make great demands on those who must assess the work and maintain a level playing field for any given student cohort. The challenge for design education today is to develop such student independence while coping with large and diverse student groups. Tailored learning strategies must be part of this involving focused application of peer support, a greater use of design professionals and well-made self-study resources – all facilitated through online learning environments to create an effective blended learning model.

2. MOTIVATION TO LEARN IS PROMOTED WHEN THE KNOWLEDGE TO BE LEARNED IS PERCEIVED TO BE MEANINGFULLY RELATED TO A LEARNER'S GOALS

Clearly one of the reason many students follow a degree course in design is to achieve their 'passport' to one of the design or design-related professions. If the passport only took the form of a paper qualification we might rightly view it as an example of extrinsic motivation. But as Tovey (2012) has shown, such a passport is also a portfolio of skills, knowledge and familiarity with working practices that are held to be relevant for professional working. Achievement of this passport makes such great demands of a student of design that a large measure of intrinsic motivation must be present to succeed. The reason design students display such intrinsic motivation might be because they find the creative process stimulating, but equally influential has been the freedom traditionally given to design students to define a brief and how they might resolve it. The authenticity of a design task can be given more weight through the participation of sponsors or clients from the professional world. Good design teaching recognises the need to establish connections between the instructional environment, including content, teaching strategies and social organisation, and the learner's goals, learning styles and past experiences. The aim must be to improve relevance of the learning experiences amidst constant change in the professional practices they seek to emulate.

Perhaps it's time for a radical rethink in our strategies for helping students to achieve their passport to the particular branch of the design profession they are aiming for. Perhaps passport development should become the responsibility of the design professions with students opting into particular university-based modules for certain skills and knowledge. This turns the current model on its head. From the start the participants wouldn't be students but junior designers. Companies would be expected to have a strategy for design teaching and learning as part of normal staff development. Central government finance that currently follows students to universities would instead be channelled to participating members of the private sector. There have been small-scale overtures towards this sort of model from organisations such as Dyson. Clearly there are major logistical problems in placing all students under this new umbrella. It could be brutal for all but the most committed junior designers and there might not be a place for those who want to develop design skills but then transfer and apply them outside of design. But as far as this principle is concerned it might offer a radical new way to make learning meaningful.

3. MOTIVATION TO LEARN IS PROMOTED WHEN LEARNERS BELIEVE THEY CAN SUCCEED IN MASTERING THE LEARNING TASK

One would think that self-belief might be in short supply in design education. Not only are students expected to cope with a wide range of experiences but they also have to take risks and work with ill-defined problems where there is no one single correct answer. Where are they to get their confidence from? The answer lies in constructing a curriculum that offers plenty of opportunities for personal control and fosters an expectancy for success. As Keller points out 'Confidence is achieved by helping students build positive expectancies for success and then experience success under conditions where they attribute their accomplishments to their own abilities and efforts rather than to external factors such as luck or task difficulty.' Confidence is partly promoted through dialogues – within the peer group, with other students, with tutors – but these are becoming ever more difficult to foster given the large class sizes.

The notion of 'mastering the learning task' can give the false impression that there is someone (the tutor) who is not only in control but fully understands the point the learner is trying to get to. In reality most tutors and students travel together on the learning journey. This is particularly true in the teaching and learning of design where conjecture, in the form of drawings, digital models and physical constructions, assists the verbal communication towards a shared understanding. It's not one-way traffic. Both learner and teacher use conjecture to test their own thinking at the same time as communicating. It gives rise to serendipitous discoveries or changes in direction and it can be genuinely cooperative. It's active learning and a key responsibility for new models of design education is to help students learn to learn. As Keller notes, mastering learning provides a key foundation for confidence and motivation.

4. MOTIVATION TO LEARN IS PROMOTED WHEN LEARNERS ANTICIPATE AND EXPERIENCE SATISFYING OUTCOMES TO A LEARNING TASK

Few people will have escaped the notorious 'satisfaction surveys'. They seem ubiquitous in today's world of retailing, health care and any number of branches of our consumer culture and education has not escaped. What such satisfaction surveys in education should be trying to achieve is to ascertain to what extent learners have positive feelings about their learning experiences such that they can develop continuing motivation to learn. However, such surveys are often misdirected or misused. There are two priorities here: first is to create learning environments that give rise to satisfaction and second to apply relevant tools to assess the level of satisfaction. Both need some attention if they are to support the co-working of extrinsic and intrinsic motivational forces.

Design activity is challenging but it also offers students a safe environment to engage in the taking of risks. Both challenge and risk can be the source of personal satisfaction but often they need to be tailored to suit the particular needs of individual learners or small teams. Similarly conflict, particularly overcoming conflict, can be stimulating and satisfying giving rise to enhanced confidence and motivation. Anxiety about failure is one of the most common inhibitors to good academic performance but the nature of design means that anxiety and failure needn't be linked. It was Tim Brown, CEO of design consultancy IDEO, who said that the purpose of design activity was to 'fail faster' (Brown, 2008). By this he meant that designers should seek out problems with their ideas through various strategies of modelling and prototyping in order to discover the failings as quickly as possible. In contrast to a strategy of 'selling' ones ideas only to have others identify problems later down the development line, Brown insisted that failing faster offered a better route to the best possible design, in the shortest time and the lowest cost. There is a need to unpack failure and to let an understanding of failure guide assessment. As designers we need to be hungry for failure!

Of course, encouraging and coping with failure is easier said than done. Inevitably, failure occurs when there is limited time to rethink, remake, re-present. The temptation is to sell an idea rather than test it. But it's just that ability to reflect on a particular design proposal that marks a maturity in design thinking that is central to the passport to practice. Achieving satisfaction from being able to identify and distinguish between those constituent parts of a design idea that are worth keeping and those that are failing is an indicator of mature design thinking and underpins motivation to learn.

Motivation is enhanced through providing students with opportunities to apply what they have learned. When this is coupled with personal recognition, attention to overloading and a fair and equitable playing field a virtuous circle is created feeding intrinsic feelings of satisfaction.

5. MOTIVATION TO LEARN IS PROMOTED AND MAINTAINED WHEN LEARNERS EMPLOY VOLITIONAL (SELF-REGULATORY) STRATEGIES TO PROTECT THEIR INTENTIONS

This fifth principle defined by Keller refers to the necessity for a learner to be able to maintain motivation towards a goal. It's all very well exploiting strategies that stimulate motivation but unless they embody some form of stickability then the motivation is in danger of receding over time. So this fifth principle concerns persistence. Why might motivation fall away? Because there are all sorts of what Keller terms 'distractions, obstacles and competing goals'. In the world of design there might be social, cultural or technological changes that mean a chosen direction for resolving a design problem is now longer relevant. For example, printed newspapers have become less relevant as news has become a more user-generated service industry, with multiple online access points. Also legislation and protection can make the long haul of product development less than attractive or seem less commercially viable. So motivation is not just a case of being confident or finding a task meaningful etc., as suggested by the preceding four principles, the learner, and in this case the design learner, needs to find strategies to maintain their intellectual and physical energy and interest. Partly, the studio system of a co-located network of learners went some way towards supporting this. It continues to offer a way of reinforcing shared interests and can act as a buffer against setbacks though peer-group relationships. But achieving effective online or virtual studios has proved more difficult. Nevertheless, whether proximal or virtual, a shared appreciation of the power of motivation can enable groups to support each other, the dialogue being reinforced through the language of representations within which lie opportunities for reinforcing confidence or sharing concerns. Even simply being aware that motivation erodes with time can be sufficient for learners to be on guard for discouragement and to engage in strategies to bolster positive thinking.

Learning and Teaching Motivation in Design

In the introduction we suggested that design curricula in our universities currently place great emphasis on developing skills and knowledge while neglecting the development of motivation for operating across the landscape of today's and tomorrow's design practice. Perhaps there is an assumption that students have the same drive and enthusiasm as their tutors, or perhaps that a hunger for success in the form of assignment grades or career opportunities is sufficient motivation. Clive Mockford, Managing Director of Engineering Creatives, highlights the same misconception in professional practice:

> You might imagine that most designers are motivated by money but in my experience the most powerful rewards are those associated with being part of successful innovation; working as part of a team to successfully get a product into the marketplace where it's well received. It's an emotional reward rather than a financial reward. This is another reason why it's hard for undergraduate design courses to give students realistic experiences – the assessments create false reward systems. (Mockingford, 2009)

There is frequently a rigorous search for enthusiasm at the point of entry onto design degrees, which is why many universities still insist on interviewing design candidates, but once admitted to a degree programme in one of the many and varied design fields, the development of motivation is left largely to the individual. However, motivation is not some optional add-on. It has a function equal in significance to the whole landscape of skills and knowledge. Motivation is not a passive force; it can be understood, shaped and developed. In fact, designing demands a constant refreshing and renewal of motivation, and this has implications for design curricula and the sort of learning experiences that are created for students.

Motivation is a vital force in design thinking. It underpins design action and design learning. We emphasise design learning because all designing is about learning, whether in professional practice or education. This emphasis was characterised by Garner and Evans (2012) as the 'learningscape' of design and it was proposed that a key requirement for navigation around such a learningscape was motivation. Partly it provides a form of energy to the learner. This energy might manifest itself as enthusiasm to engage with chosen skills or knowledge or curiosity to explore unknown or undeveloped skills and knowledge. It is also an energy to persevere with skills and knowledge when their value is not clear or their application is without direction. Given that the value of motivation is widely accepted, it seems strange that strategies for developing motivation are so thin on the ground.

Motivation is enhanced by challenge and intellectual risk, but the nature of the challenges and risk placed before design students can be too closely controlled. The reasons for this control are understandable: the demands of closely defined curricula, the requirement to focus on particular skills and knowledge, the pressure to assess students against particular learning outcomes and the need to offer consistent experiences to large student cohorts have combined to impoverish the nature of the challenge and risk. Compared to the challenge and risk in professional practice, the preparation on degree programmes can seem limp. Challenge needs to be supported by the development of strategy.

Many designers display a natural ability to exploit simplifications to make complicated tasks understandable and manageable. After all, design models are no more than simplifications to assist the understanding of problems or the communication or testing of ideas. Feeling able to manage is particularly important where design tasks present ambiguity or conflict between forces. Motivation is enhanced if we feel we have a strategy for dealing with the complexity or ambiguity of tasks. This might be something simple such as breaking a job down into manageable chunks or into a sequence, but it also includes understanding the processes that have proved effective for other people. Also motivation relies on an ability for self-evaluation. If an individual is unable to measure their own performance in relation to that of others, then they learn to rely on the opinions and judgements of peers and tutors. Independence and well-founded resilience are vital in order to cope with conflict and confidently defend proposals. Traditionally, universities have not been very good at developing students' ability for evaluating their own performance, but self-evaluation is not only empowering, it has become a key skill for employment in today's creative industries.

Motivated Professional Designers

The design professions, ranging from communication design to urban planning, face radically new responsibilities concerning their ability to act as interpreters and transformers of conflict. The design professions already look for new graduates who can operate confidently at the interface between opposing forces, and the skills and knowledge required are set to increase. But skills and knowledge without motivation are little use for anything other than routine design. In situations that demand design leadership, motivation can be more valuable than skills and knowledge.

This chapter has made a case for prioritising the development of motivation in young designers. This is all the more urgent as forces conspire to erode motivation by swamping design tasks with information. Designing is a learning task as much as a creating task, so designers need support for the agile navigation of the world of design. We need learning experiences that tap into students' natural motivations but which professionalise motivation to create a resilient, informed and sustainable capacity. Since motivation is not one distinct force but is shaped and coloured by numerous cognitive forces and emotions, it seems logical that any attempt to develop motivation should acknowledge its diversity. For example it might be necessary to work on curiosity through investigative tasks. These might tap into a designer's emerging ability to communicate. Other approaches to developing motivation might focus on attitudes or beliefs, personal interests or the nature of obsession.

There exists a form of design intelligence – an exercising of design thinking skills – that is well-suited to addressing some of the conflicts in today's world. Design intelligence recognises the ubiquity of designing, the role of the past in the creation of the future and the potential for design to be a problem-causing phenomenon as well as a problem-solving one. Design intelligence helps us address the what, where, when, how, who and why questions of design, but it is a capacity built on potentially fragile foundations of motivation, attitudes, values and beliefs. Design intelligence isn't robust and resilient; it can be hesitant and fragile, but it is essentially optimistic, inclusive, reflective and creative in the way it frames and seeks to resolve conflict.

References

Brown, T. (2008) Design thinking, *Harvard Business Review*, June, available at http://www.ideo.com/images/uploads/thoughts/IDEO_HBR_Design_Thinking.pdf.

Chickering, A.W. and Gamson, Z.F. (1986) Seven principles for good practice in undergraduate education, *Washington Center News*, Fall. http://www.lonestar.edu/multimedia/SevenPrinciples.pdf.

Garner, S. and Evans C. (2012) The learningscape of design, in S. Garner and C. Evans (eds), *Design and Designing, A Critical Introduction*, pp. 444–458. London: Berg.

Keller, J.M. (1979). Motivation and instructional design: a theoretical perspective. *Journal of Instructional Development*, 2(4), 26–34.

Keller, J.M. (2008) First principles of motivation to learn and e3/learning, *Distance Education*, 29(2), 175–185.

Liu, E.Z.F., Lin, C.H. *et al.* (2012) The dynamics of motivation and learning strategy in a creativity-supporting learning environment in higher education, *Turkish Online Journal of Educational Technology*, 11(1).

Mockford, C. (2009) Unpublished interview with Clive Mockford, Managing Director, Engineering Creatives, 24 March.

Sternberg, R.J. and Lubart, T.I. (1991) An investment of creativity and its development, *Human Development*, 34, 1–31.

Tovey, M. (2012) The passport to practice, in S. Garner and C. Evans (eds), *Design and Designing, A Critical Introduction*, pp. 5–19. London: Berg.

Woodman, R.W. and Schoenfeldt, L.F. (1990) An interactionist model of creative behavior. *Journal of Creative Behavior*, 24, 279–290.

Chapter 5
Signature Pedagogies in Design

ALISON SHREEVE

What are Signature Pedagogies?

The idea of signature pedagogies was developed by Shulman (2005a, 2005b) at the Carnegie Foundation for the Advancement of Teaching in the United States of America to explain disciplinary differences in the way that some subjects were taught and how students were supported to learn. He identified that learning to become a professional, such as a doctor, lawyer or clergyman, each had particular ways in which teaching was structured, which were different and distinctive enough to be associated only with learning in that particular profession. These 'signature pedagogies' were used because they helped students to think and behave like a professional, in order to practice:

> One might therefore say that professional education is about developing pedagogies to link ideas, practices, and values under conditions of inherent uncertainty that necessitate not only judgment in order to act, but also cognizance of the consequences of one's action. In the presence of uncertainty, one is obligated to learn from experience. (Shulman, 2005b: 19)

The identification of signature pedagogies helps to explain that learning to become a design practitioner is not simply a matter of knowing facts, but is a much richer and deeper experience which requires a change in knowledge, behaviours and emotion. It moves the emphasis away from content of the curriculum and explains the importance of practical, embodied and experiential ways of knowing and being which are essential to the profession or practice the learner is intending to enter. Such explanations of learning have been described as an ontological approach (Dall'Alba and Barnycle, 2007; Adams et al., 2011), which requires a change in the person, or a whole-person approach to learning, rather than an epistemological approach which concentrates on the knowledge, and usually only the explicit knowledge, that can be listed in the curriculum. Drew (2004) identified that lecturers in art and design tended to adopt a community of practice (Lave and Wenger, 1991; Wenger, 1998) approach to their teaching; they were more likely than other subject areas to be trying to develop their students as practitioners with an emphasis on the practice, or profession, which they were hoping to enter on graduation. Tutors adopting this approach to teaching were focused on enabling their students to become practitioners rather than reproducing facts. It could be argued that students were being helped to become members of a community of practice (Wenger, 1998) by the 'old timers' or experienced members of the community. There is therefore supporting research to suggest that signature pedagogies do exist in art and design disciplines, because an approach which identifies teaching as inculcating students into a community of practice requires teachers to devise learning as a gradual enculturation into the languages and the practices of the discipline. Signature pedagogies are learning activities that help students to think and act like design professionals.

Why is it Important to Identify Signature Pedagogies?

An awareness of our own pedagogic traditions helps us to understand, but also to critique how we support students to learn. Knowing that there are specific ways to help students be prepared for life beyond university is important, and in an increasingly uncertain world it is becoming more so. In many Western tertiary education systems students are investing increasingly large sums of money in their education and are becoming rightly more critical, aware and demanding. If we are able to explain why we use particular learning approaches this is helpful to students, it encourages them to understand their own learning journey and to see how it prepares them for the future. It enables students to see how they have progressed and to have confidence in their learning and confidence in their tutors. This is particularly important where there is very little in the way of an explicit syllabus and learning is an unknown adventure with uncertain outcomes depending on students' individual creative paths and responses. Both student and tutor can find themselves in positions of ambiguity, when creative experimentation breaks new and unforeseen territories (Austerlitz et al., 2008).

For tutors it is important to maintain, change and evolve the signature pedagogies of their particular discipline. As practices change beyond the academy they need to ensure that the way learning activities are designed continues to prepare students for the changing world. In this way tutors help student employability, their readiness to contribute to the economy and their ability to understand the opportunities available on graduation. Signature pedagogies can therefore be mutable, specific to a design discipline, and also generic across design disciplines. Whether we continue to use certain approaches and techniques for learning should be dependent on continuing validity; does this way of learning support my students to think and behave like a certain kind of designer able to respond to the issues of the economy and to communities?

Signature pedagogies for design are those ways of learning which help students to become designers; to think and act in ways which are deemed to be professional and appropriate. As there are many different disciplines within design it would be unusual if they didn't have a range of different signature pedagogies. This chapter will draw on empirical work across different disciplines (Sims, 2008) and from observations and conversations with colleagues to illustrate both generic and specific signature pedagogies. It does not attempt to be all-inclusive or to identify best practice, simply to illustrate how specific pedagogic practices could be argued to contribute towards thinking and behaving like a design practitioner.

The Studio

The site of learning could in itself be seen as a signature pedagogy. Schön (1985) identified in his seminal work that the studio was a site of particular cultural practices which included ways of teaching and modes of being and acting. Smith-Taylor (2009) claims that the studio constructs particular ways of teaching that lead to a student-centred approach (Prosser and Trigwell, 1991) as it removes the tutor from the centre of focus and requires them to be mobile, working in different parts of the studio and engaging students in dialogue. Students who do have a permanent or semi-permanent location for their learning are supported to learn through peer engagement and where different levels of experience are co-taught students can be supported to envisage where their learning will take them through their degree programme. They have access to peers with more experience who are closer to them in the learning journey than their tutors are and this is typical of a community of practice (Wenger, 1998), where designers know and often work with other designers, whether in a physical space or not. The design community has its own culture, networks and languages, or in Wenger's terms a community of practice, and students are on the edge of the community, with the intention of becoming a main part of the community. They are therefore being prepared to think and behave like a designer in more than one way through their physical learning environment.

With increased financial pressure on many design educational institutions (Clarke and Budge, 2010) students may no longer have a studio to themselves with a permanent work station. However, a shared learning environment is usually created, which continues to offer a studio-based ethos for learning. In an online community similar exchanges, participation, articulation and discussion mirrors the physical studio environment and can support the engagement and practising of disciplinary languages, including visual, written and spoken forms. In some cases, the online learning environment is a more appropriate way to communicate, as it more closely replicates the situation that professionals will be using. A study of the online community of learners in a postgraduate photojournalism course (Lowe and Blythman, 2009) indicated that this dispersed learning community, who were by necessity spread around the world, also developed and engaged with professionals who were in a similar situation. Thus the engagement of the professional and learning communities online helped to develop the learners' understanding of becoming a professional photojournalist. This is a good example of how technologies are able to create a new kind of 'studio' where shared experiences are essential to learning and the boundaries between professional practitioner and student become blurred.

Projects and the Brief

Most design courses will use a project-based, experiential learning approach with students. This was identified by de la Harpe and Peterson (2008) as one of the main teaching mechanisms used in design disciplines, along with the critique and making. The project is a way to provide a focus for learning, but in doing so is usually presented through a written 'brief', in a way which reflects the complexity of design opportunities outside the academy. There will be a fixed time to complete the work, a requirement specifying format, restraints and parameters for the design work and its presentation. Students will be required to undertake 'research', which is often left as a tacitly understood process and expectation (Kjolberg, 2013), which I will return to later in the chapter. The outcomes of learning, in terms of the finished design or artefact are open-ended and largely unknown to the tutor at the time of writing the brief. This is an essential factor in developing a creative response from the student. They are expected to take an autonomous approach to finding their own way through the brief, which can be a very unsettling experience for the less confident student. In some cases this is also an unsettling experience for the tutor, who guides and advises, rather than dictates to the student, as the object of the learning is for the student to use their own creativity to explore and provide an 'answer'. Radical journeys of discovery by the student may go way beyond a tutor's experience and this presents risks for both parties. However, this extreme is where new knowledge can be created in design and requires an act of faith for both, but it helps to contribute to a pedagogy of uncertainty (Austerlitz et al., 2008) which characterises both learning and professional worlds, where creative responses are essential.

Materiality

The project is a vehicle for 'doing and making', a pedagogic approach which is at the core of most learning experiences in design. The value of this experiential knowledge is still at the heart of many disciplines, even those primarily working in a digital format. For example, a graphics tutor will make every effort to ensure that his students experience typeface and printing as a physical activity, in order to understand type as embodied knowledge. He needs his students to have a physical experience of proportion, letter construction and spacing. Although design may professionally be carried out on a computer, printed artefacts still require an understanding of ink, paper, printing processes and scale. Even experiencing how paper responds to ink and the ensuing cultural signification of such material artefacts is part of the knowledge of a practising designer and one of the reasons why many tutors will still offer a practical, hands-on approach to the physical and material world of their design discipline.

Materiality, whether artefacts are designed on a computer, made by a computer or in more traditional ways, is a fundamental aspect of learning and could be considered to be part of the signature pedagogy of 'doing and making'. Learning manifests itself in something: an artefact, a performance, a design, a portfolio. This trace or evidence of a learning process is central to other learning activities. Because the evidence exists in a material form which can be subject to scrutiny, objects and processes of designing are central to debates, discussion and questioning. This leads to two other signature pedagogies, the critique and the ongoing dialogue which supports learning.

Dialogue

In a study of the learning environment (Sims 2008) in four disciplines in art and design, one tutor characterised teaching as 'a kind of exchange' (see also Shreeve et al.,2008). This is indicative of the dialogic nature of learning, where tutors are probing, prompting and questioning in order to encourage students to make decisions, to consider possibilities and alternatives, before deciding on a course of action in their project work. This 'exchange' is the fundamental pedagogic practice which both models the way that designers think and also encourages students to practice it. Dialogue makes visible, or audible, what normally remains tacitly understood in the realm of design practice. When tutors are teaching it may sound like a conversation. Where large groups of students are taught together, this dialogue is harder to maintain and increasingly pressure to deliver to more and more students will require alternative ways to enable dialogue, whether through peer learning, small group activity, the inclusion of practising designers in projects (live projects) or through placement and internships, students need to learn how designers think by learning the language (Logan, 2006) and reflective practices of design. Dialogue characterises the student-centred approach of design disciplines and is fundamental to engagement of students, enabling them to practice arguments, explain thinking processes and learn the languages of design, whether verbal, visual, critical, historic or contemporary. For many tutors the critique (crit) is still the focus of group discussion about work, either as work in progress with formative feedback, or as summative assessment.

The Crit

The critique was the first signature pedagogy to be identified in the literature (Klebsedel and Kornetsky, 2009). Much research and debate has taken place over the last fifteen years in regard to the crit (for example Blair, 2007; Percy, 2004; Webster, 2006). Although the critique itself has probably changed over the years and has hopefully become less of a traumatic event for students, it still remains one of the primary ways to establish appropriate design standards of thinking, evaluation and response to a project brief. It has been noted that the practice of the crit itself is seldom carried out in professional design studios, although there may be critical discussion around ideas and designs. The pedagogic function of the crit should be to establish standards, to provide positive feedback to students on performance and to share alternative perspectives on design possibilities, demonstrating to students that success is possible in a number of ways, that there is no right answer. It also models the thought processes, critical analysis and language needed to become a designer.

The debate around the efficacy of the crit as a teaching practice is important, as it highlights the fact that a signature pedagogy is not necessarily best practice, only that it has become a significant method in the tutor's repertoire. Positive changes to the crit have made the format a more participatory one, rather than the instructive tutor-centred performance that many remember from their own student days. The variations on the crit still retain the distinctive pedagogy, a review of progress and achievement carried out as a social activity, a practice which is possibly rare in most other disciplines.

Research as a Signature Pedagogy?

The requirement for undergraduate students to 'research' in order to proceed with project work is extensive in design. However, one could argue that there is often no associated pedagogic practice which enables students to learn to do research (Kjolberg, 2013), other than a requirement to go away and 'do research'. I have however included research here as part of the signature pedagogies identified with Ellen Sims (Sims 2008, Shreeve, Wareing and Drew, 2008) because undergraduate research is an increasingly widely used pedagogic approach across other disciplines such as geography and the sciences (Healey and Jenkins, 2009). Reading about other disciplinary approaches which use undergraduates on 'live' research projects is interesting as it illuminates pedagogic approaches which are taken for granted as part of 'what we do', the everyday practice of learning and teaching in art and design. Students are expected to 'research', that is, explore the context of the brief, their own response to it and to possible factors which might influence decision-making. This may include primary research with end users or secondary research (around the impact of process and materials on the environment for example) but it also includes the process of ideas generation and development for designs. The expectation of 'doing research' is probably fairly universal in design, but the pedagogic methods for supporting students to understand and develop a process for it is perhaps less well formed. A study by Shreeve et al. (2003) indicates that fashion and textiles students can approach the research component of their projects with different intentions and different strategies. This includes a focus on reproducing visual elements of the world, through to more inclusive approaches which focus on creating personal conceptual responses to a design brief with a more holistic interpretation and personal, individual construct.

The pedagogies which support students' learning through the research component of their project would be fruitful to explore in more detail. However, modified approaches to Edward de Bono's lateral thinking strategies have been successfully used in graphics (Raven and Smith, 2007) and many tutors have devised their own workshops which concentrate on the ability to analyse aspects required for generating new designs, such as an ability to analyse visual objects and materials for example. Research is therefore perhaps better considered as a component of learning through project work and through material culture, rather than a signature pedagogy. Ellen Sims and I included it in our original work on signature pedagogies because it was common to all the disciplines we studied across four different colleges. The actual pedagogies associated with students' research are, however, rather a murky area as Kjolberg has shown.

Differing Signature Pedagogies

The pedagogic approaches discussed above have referred to aspects which are fairly common in UK design education, but as the changing nature of the crit and the technological impact of distance learning and online learning indicate, signature pedagogies are mutable. If they are defined as those aspects of learning which help students to think and behave like professionals, as the design professions change, so too will the signature pedagogies.

Although generic approaches have been identified above, I will illustrate that for some design disciplines there are other ways of teaching which have evolved or been required to enable students to enter into the community of practice as a novice, before they actually enter the profession on graduation.

The role of a creative advertiser is to generate ideas and scenarios to promote products and services. In order to do this a tutor will use workshops to role play, push the boundaries of the acceptable and challenge what is 'right'. Students are required to develop ways of thinking which challenge the norms as well as evolve them, and are encouraged to approach situations from oblique angles. As the advertising industry employs creative teams students often work in pairs and develop a joint working arrangement which can lead to a lifetime working together, often setting up their own creative agency. Thus working in pairs or small teams would fulfil Shulman's definition of a signature pedagogy as the habit of mind developed through partnerships continues into the professional practice of the advertising industry.

For textiles students the opportunity to present their work at an international trade fair in Paris, *Indigo,* is a practice which crosses the boundary between education and work. Where participation in the trade fair is part of the students' learning journey they are supported to understand how such trade fairs are organised and how central to their future career they may be, as this is where many textile designs will be sold. A project at Chelsea School of Art enabled students to carry out, with tutor supervision, all the aspects of planning, organisation, transport, design of the trade stand, presentation and selling. Students who are able to take part in such live events gain a much more rounded view of the work involved as a professional designer and learn to appreciate the hidden skills professionals need. One student in this project said that she learned more about herself through doing the project and learned too that she enjoyed selling even more than designing. Opportunities to participate as equals in professional spheres of activity are increasingly important and are arguably examples of signature pedagogies because they are exactly replicating professional practices or including students in professional worlds. The importance of these kinds of participation is the need to identify the learning opportunities for students, rather than the tutor organising everything. If students are enabled to see how to prepare, plan and organise participation in an exhibition, trade show or sales arena they are better prepared for the professional role of designer. If they are explicitly shown how participation in a live project is fundamental to learning ways of being as a design professional they are better able to understand the value of this as an educational experience relevant to their future employment.

Conclusion

Signature pedagogies are a distinctive and interesting phenomenon to explore in design education because they are directly linked to the professional practices beyond the university and the knowledge, skills, behaviours and understanding students will need as designers. They help to bridge two worlds or cultures, that of academia and the professions. By exploring changes in design practice this can help to identify new approaches to learning and teaching in design and encourage a creative approach to the structure and content of the curriculum and its learning activities. In a growing climate of economic uncertainty it may be harder to maintain the dialogic exchanges which are so important to individual student development, but through examining the work of professional designers new ways to learn how to think and behave like a professional offer up hope that we can continue to maintain the high standards of the UK's design education. Signature pedagogies will continuously evolve and the concept of learning and teaching practices which can support students to become creative design professionals should be at the core of every tutor's intentions when preparing learning activities for students.

References

Adams, R.S., Daly, S.R., Mann, L.W., Dall'Alba, G. (2011) Being a professional: three lenses into design thinking, acting, and being, *Design Studies*, 32(6), 588–607.

Austerlitz, N., Blythman, M., Grove-White, A., Jones, B.A., Jones, C.A., Morgan, S., Orr, S., Shreeve, A. and Vaughan, S. (2008) Mind the gap: expectations, ambiguity and pedagogy within art and design higher education, in L. Drew (ed.), *The Student Experience in Art and Design Higher Education: Drivers for Change*. Cambridge: JRA Publishing.

Blair, B. (2007) At the end of a huge crit in the summer, it was crap – i'd worked really hard but all she said was 'fine' and i was gutted, *Art, Design and Communication in Higher Education*, 5(2), 83–95.

Clarke, A. and Budge, K. (2010) Listening for creative voices amid the cacophony of fiscal complaint about art and design education, *International Journal of Art and Design Education*, 29(2), 153–162.

Dall'Alba, G. and Barnycle, R. (2007) An ontological turn for higher education, *Studies in Higher Education*, 32(6).

de la Harpe, B. and Peterson, F. (2008) Through the looking glass: what do academics in art, design and architecture publish about most?, *Art, Design and Communication in Higher Education*, 7(3), 135–154.

Drew, L. (2004) The experience of teaching creative practices: conceptions and approaches to teaching in the community of practice dimension, in A. Davies (ed.), *Enhancing the Curricula: Towards the Scholarship of Teaching in Art, Design and Communication, Proceedings of the Second International Conference*, pp. 106–123. London: Centre for Learning and Teaching in Art and Design.

Healey, M. and Jenkins, A. (2009) *Developing Undergraduate Research and Inquiry*. Higher Education Academy. York, UK.

Kjolberg, T. (2013) Visual Research in Fashion and Textile Design Undergraduate Education. Unpublished Ph.D. Thesis, University of Brighton.

Klebsedel, H. and Kornetsky, L. (2009) Critique as signature pedagogy in the arts, in R.A. Gurung, N.L. Chick and A. Haynie (eds), *Exploring Signature Pedagogies: Approaches to Teaching Disciplinary Habits of Mind.* Sterling, VA: Stylus.

Lave, J. and Wenger, E. (1991) *Situated Learning: Legitimate Peripheral Participation.* Cambridge: Cambridge University Press.

Logan, C. (2006) Circles of practice: educational and professional graphic design, *Journal of Workplace Learning,* 18(6), 331–343.

Lowe, P. and Blythman, M. (2009) Blogs and the reflective practitioner: professional not confessional, in J. O'Donoghue, *Technology-supported Environments for Personalized Learning: Methods and Case Studies,* pp. 149–166. Hershey, PA: Information Science Publishing.

Percy, C. (2004) Critical absence versus critical engagement: problematics of the crit in design learning and teaching, *Art, Design and Communication in Higher Education,* 2(3), 143–154.

Prosser, M. and Trigwell, K. (1991) *Understanding Learning and Teaching.* Buckingham: SRHE/Open University Press.

Schön, D.A. (1985) *The Design Studio: An Exploration of Its Traditions and Potentials.* London: RIBA Publications Ltd.

Raven, D. and Smith, C. (2007) *Thinking Tools to Encourage Creative Learning.* Available from http://www.arts.ac.uk/librarylearningandteaching/clipcetl/academicpapers/papers2007/ last accessed July 2013.

Shreeve, A., Bailey, S. and Drew, L. (2003) Students' approaches to the 'research' component in the fashion design project, *Art, Design and Communication in Higher Education,* 2(3), 113–130.

Shreeve, A., Wareing, S. and Drew, L. (2008) Key aspects of teaching and learning in the visual arts, in H. Fry, S. Ketteridge and S. Marshall (eds), *A Handbook of Learning in Higher Education,* 3rd edn. London: Kogan Page,

Shulman, L.S. (2005a) Signature pedagogies in the professions, *Daedelus*, 134(3), 52–59.

Shulman, L.S. (2005b) Pedagogies of uncertainty, *Liberal Education*, 91, 18–25.

Sims, E. (2008) *Teaching Landscapes in Creative Arts Subjects. Report on the CLIP CETL Funded UAL Research Project*. Retrieved 29 September 2010 from http://www.arts.ac.uk/docs/Landscapes-final-report.pdf.

Smith Taylor, S. (2009) Effects of studio space on teaching and learning: preliminary findings from two case studies, *Innovative Higher Education*, 33(4), 217–228.

Webster, H. (2006) A Foucauldian look at the design jury, *Art Design and Communication in Higher Education*, 5(1), 5–19.

Wenger, E. (1998) *Communities of Practice. Learning Meaning and Identity*. Cambridge: Cambridge University Press.

Chapter 6

The Experience of Teaching a Creative Practice:
An Exploration of Conceptions and Approaches to Teaching, Linking Variation and the Community of Practice

LINDA DREW

Introduction

This chapter focuses on the research I completed nearly a decade ago, which stemmed from a desire to understand why learning and teaching in design is different, or demonstrates variation from, most other university disciplines (Drew, 2004). Or so I thought. In my quest to understand learning and teaching in design, I discovered that I should explore the teaching of creative practice, that is to say, the teaching of subjects that are framed by the concept of becoming a creative practitioner. The quest involved research into conceptions and approaches to teaching and also the concept of the community of practice (Wenger, 1998).

Much of the practical description of approaches to teaching in this context had not been systematically attempted in the last decade. A good summary of descriptions of the context, approaches and case study material can now be found in the most recent textbook on teaching in this context (Shreeve, Wareing and Drew, 2008).

Most studies of conceptions and approaches to teaching have chosen to focus on traditional university subject disciplines, e.g. sciences and humanities. There are however some earlier studies of conceptions of teaching in creative practice-based disciplines including music (Reid, 2000) and design (Drew, 2000a, 2000b).

In that study of design teachers five qualitatively different conceptions of design teaching are described (Drew, 2000b). These range from the teacher offering a range of practical and technical skills to students, through to the teaching as helping to change students' conceptions. These categories of description illustrate a dimension of the qualitative variation in design teaching. Conception A in this study illustrates that even one-to-one teaching contexts can still be conceptualised in transmission terms. The other four conceptions identified incorporate a degree of student-centredness which increases from B to E. Categories D and E also demonstrate a community of practice dimension as a focus for the context of teaching.

The practice-based context of studio teaching can be seen as a student-centred approach, but as Reid and Davies (2000) report, some teachers in this context hold conceptions of teaching as instructional and teacher-focused rather than co-operative and collaborative learning. The quality of this learning environment relates to the context and the conceptions of learning and teaching held by both teacher and student (Reid and Davies, 2000).

Trigwell, Prosser and Taylor (1994) reported five approaches to teaching using transcripts from the interviews with 24 university physics and chemistry teachers from an earlier study to conduct their analysis. The analysis of categories of approaches to teaching was conducted in terms of the strategies teachers adopted and the intentions which informed those strategies.

Prosser and Trigwell used the outcomes of these qualitative studies to devise the *Approaches to Teaching Inventory* (Trigwell and Prosser, 1996a). Use of the questionnaire confirmed the relationship between intention and strategy which was found previously. The Approaches to Teaching Inventory or ATI was developed to measure the variation in the ways teachers approach their teaching in a particular situation as it is a relational instrument. The teaching of art and design subjects is often described in ways that are quite different to descriptions of teaching in more traditional subject areas such as science. Less use is made of lecturing and lecture notes, the activities that are employed tend to be more studio- and project-based, and involve smaller groups of students than in the more traditional areas.

Studies which embrace the sociocultural perspective on practice particularly emphasise learning to practice in various settings. Learning to practice, whether in the workplace or simulated settings, is seen as a move towards full participation in a community of practice (Lave and Wenger, 1991; Lave, 1993). That move to full participation takes place by engaging in 'legitimate peripheral participation' which is taking part in the authentic activities of the practice, albeit with guidance, and at the edges of the community. These views emphasise social practice as a premise for learning and that 'knowing in practice' arises from participation in that social practice (Billett, 1998).

Learning that results from participation in social practices means that the participants appropriate ways of seeing the world inherent in those practices. These situational and social factors are a key part of learning to practice (Billett, 2001). Billett argues that a non-dualist view of learning is becoming more accepted, based on the concept that there is an inseparable relationship between an individual's knowing and their social life-world (Rogoff, 1990). Many would argue that preparing learners for life as a creative practitioner, be that as an artist or a photographer, is essentially preparing them for solitary work. Rogoff (1990) suggests that cultural practices and norms shape even the most apparently solitary activities. This is further confirmed by Billett:

> An artist working in the isolation of his studio reported shaping his practice to account for situational factors determining the kinds and purposes of his work that included physical environments and consideration of the market. (2001: 444)

Jean Lave describes the social participatory perspective on learning as individuals developing and changing their identities, 'people are becoming kinds of persons' (Lave, 1996: 157). Lave's study of the apprenticeship of tailors in Liberia during the 1970s identifies how the tailors were primarily making ready-to-wear trousers, but the apprentices also learned other important contextual factors about being a tailor:

> they were learning relations among the major social identities and divisions in Liberian society which they were in the business of dressing. They were learning to make a life, to make a living, to make clothes, to grow old enough, and mature enough to become master tailors, and to see the truth of the respect due to a master of their trade. (Lave, 1996: 159)

The cornerstone of these issues for professional learning can be summarised as learning to practice or becoming inducted into a community of practice

(Wenger, 1998). Wenger further defines the role of *participation* in a practice in its relationship with the *reification* of artefacts or processes particular to the practice. Wenger regards participation as 'the social experience of living in the world' (1998: 55) which involves acting, thinking and feeling as a whole personal experience. It is from participation Wenger argues, that an *identity of participation* is constituted through the relations formed in participation itself.

The ATI has been used to measure variation in approach to teaching in design teaching contexts (Trigwell, 2002). In that study it was found that, as in other teaching contexts, there is significant variation in descriptions of how teaching is approached in design subjects, and that overall, the approaches adopted by design teachers are described as being more student-focused than in most other areas of higher education teaching. The results also suggest that when design teachers describe their approaches as student-focused they are more likely to say they learn more during the teaching of their subjects and are more likely to give students the opportunity to explore their own creative ideas, than when the teaching is described in terms of teacher-focused, information transmission. The ATI was found to be an acceptable indicator of qualitative variation in teaching approaches in creative fields such as design.

Qualitative Method

The data is from an interview study of 44 teachers from 8 UK universities and is explored with a phenomenographic approach (Marton and Booth, 1997). The analysis was grouped into three discrete subdisciplines, fine art (11), design (18) and media (15), through which variation in the practice dimensions could also be discerned. Opportunity sampling was used to identify the eight university departments of art, design and communication. This paper adopts a second-order perspective on the experience of teaching a practice-based subject in art, design and communication departments.

The interviews were semi-structured and consisted of questions designed to encourage the respondent to talk about the way they perceived their teaching role and related strategies and intentions. The aim of phenomenographic analysis is to develop categories of description which illustrate the limited number of qualitatively different ways of experiencing a phenomenon, in this case the experience of teaching a practice-based subject in media, fine art and design. The categories were devised by looking for the variation between responses, and the similarities between statements within categories. Then final descriptions were produced to reflect these similarities and differences.

The descriptions of the categories were developed using two components –
how the explanation is given and what is focused on (Trigwell, 2000: 74).

The categories of description, described in the next section, are internally
related to each other. Categories were sorted into a meaningful order, with the
'lower' less complete conceptions first, moving into 'higher', more complete
conceptions. The higher conceptions encompass the lower conceptions and
are therefore more complete. This is known as a hierarchy of categories of
description, the logical relations between these categories are illustrated in the
outcome space. The outcome space is not a full rich description of teaching,
rather it is a description of those aspects of teaching that are seen to have
qualitative variation.

Conceptions of Teaching Creative Practices

For the purposes of this study, creative practice teaching is described in the
context of media, fine art and design. These subject areas include teaching the
practices of journalism, film-making, television and video production, animation
and photography, fine art, painting, sculpture, printmaking and related visual
arts, graphic design and illustration, interior design and architecture, fashion
and textile design. The most reliable and valid comparisons were available if
the discipline group was inclusive of a range of creative practices, not just pure
design topics. The constant context of each interview was the teachers' practice-
based teaching as opposed to teaching visual or historical studies for example.

CONCEPTION A: TEACHING IS OFFERING STUDENTS A RANGE OF PRACTICAL AND TECHNICAL SKILLS

The teacher aims to reinforce technical ability by giving demonstrations and
showing individual or groups of students' ways of making or doing. The
teacher believes that the students need to follow technical topics based on what
the teacher feels they need to learn. The emphasis of the learning is on a product
or artefact. The intention is to demonstrate or give examples of technical skills.

Structural aspects of this conception are concerned with the teacher's role,
in this case demonstrating, showing or instructing students how to make or
do something. There is an emphasis on correct procedures and observing or
checking that these are carried out correctly or for the students to demonstrate
some technical competence. The focus of the teaching is on technical and
practical skills. The teacher feels that they know best what skills to develop

or to teach and often refer to content or objectives of the course, rules of the practice or other practical parameters which they feel the students must master before progressing in the subject.

This teacher discusses the focus of his teaching, to demonstrate a process, observe the students practising it and check they can do it.

> First of all I will demonstrate how to correct, then we move onto something a bit more complicated, I've got these cans of Coke in the studio back there. Then I can show, if I have the camera that way they're all out, with the movement, they're all sharp. Okay that's fine, they're all sharp, But the camera's off thickness, the camera is called a female, you need to get it sharpened over there, we lose light so we have to increase the exposure and so on, just the basics, and then I've got them to do it, and I've got observation sheets which I tick off and they can demonstrate to me that they can actually do it, and I observe and tick off the observation sheet. (Media: PR13)

CONCEPTION B: TEACHING IS DEVELOPING STUDENTS' CRITICAL, PRACTICAL AND TECHNICAL SKILLS THROUGH STUDENT INTERACTION

The teacher aims to enable students to develop a critical language by working together in groups or teams to present their own work and to see the work of others. The emphasis of the learning is on peer learning and process. The teacher works with individuals, groups or teams with the intention to enable students to form opinions and ideas.

In this conception the teacher still feels it is important for students to develop practical and technical skills, but the emphasis is on learning with others, sometimes in team or group situations and often with an opportunity for critical debate.

> to encourage the way the group works, the peer group interaction is really important, for example, what student over that side of the group might have a key fabric that somebody over the side of the group might be looking for and if there are not using it then it's like, well can you give them the address of that. So it is very much dealing with practical issues and it is also reassuring them, a lot of them really do get unsure and quite worried about this module because it is such a big thing. (Design: NT2)

Teachers often describe their role as facilitating or encouraging the process of learning and of developing confidence in learners. In this conception, teachers are keen to emphasise elements of the process which actively engage with students.

CONCEPTION C: TEACHING IS DEVELOPING STUDENTS' SKILLS AND CONCEPTIONS IN THE CONTEXT OF PROFESSIONAL PRACTICE

The teacher encourages students to manage projects involving complex problem-solving skills which are set in the context of professional practice. The emphasis of the learning is on peer learning and process. The teacher works with students to develop conceptions with the intention to increase self-awareness, individual and team autonomy and for professional preparation.

In this conception teachers believe that real-world scenarios or projects as a simulation of professional practice enable high-level learning outcomes including problem-solving skills.

> So reflection, there's a kind of debate over whether journalists reflect or not, but we feel it's important for students here to play out a lot of ethical situations and scenarios and practical professional situations is a safe environment before they enter the industry, so we do encourage them to reflect, compare and contrast, look at real-life journalism is and how it compares to what they're doing. (Media: PR15)

The teacher believes that if students are brought into contact with practising subject experts they can bring a professional context to bear in relation to their work as well as developing ideas and concepts. Some teachers of fine art further described their teaching role as being an artist with students in the role of apprentices to the practice of fine art making. Apprenticeship is seen as a positive experience of being inducted into the fine art social context as well as the extension of practice and making art.

CONCEPTION D: TEACHING IS HELPING STUDENTS CHANGE CONCEPTIONS

The teacher emphasises original research and conceptual thinking skills. The emphasis of the learning is on peer learning and process. The teacher works with students with the intention to improve self-directed research, practice and conceptual skills.

The teacher feels that students should have an ability to relate key concepts to the practice, or to develop practice through critical examination of concepts or theories. Teachers in this conception also stress 'real-world' and practice-based contexts as in Conception C.

> Well, from the seminar presentation I guess that we are encouraging them to do what I was saying before, to take an area of theoretical work and to apply it to an example or a case study, and they have to learn to critically examine and reflect on that theoretical work in terms of what their thoughts are on that topic. But of course more generally, more generically, I think that they are developing their skills for research and presentation. I think that that's very important in terms of everyone doing this thing, but I mean, these are transferable skills, the ability to take a brief, and come back with a lively and animated presentation on that and to engage other people in it I think is a skill that is essential to a lot of areas in the media practices anyway. So in the seminar I think that that is an important element of it, they are developing still, or researching presentations, and working with colleagues as well. (Media: PR11)

To enable students to change conceptions of the subject, of the world and of their work is seen as an integral part of this conception. Teachers talks about expressing ideas, changing conceptions and also about learning beyond the subject boundaries as an aim for teaching in this subject.

CONCEPTION E: TEACHING IS HELPING STUDENTS TO CHANGE AS A PERSON

Teachers holding this conception again emphasise original research and conceptual thinking skills and peer learning and process. They differ from those holding Conception D in seeing teaching as a way of enabling students to change themselves as a person or to make changes in their lifeworld.

> each student comes away with having achieved something, and achieves something that takes them to another level of their existence, that's a bit ambitious but they have grown, whether or not they have learnt anything technical about photography is of less interest too me, I think they need to move on in their own lives and if they can produce a project with some collaboration with outside agencies or in the wider sense that it might be a scientific institution or a group of young mothers in an organisation, so, photography is not seen simply a means to an end,

but as a real way of shaping how people understand themselves and the world around them. (Media: FC14)

Teachers also express aspects of changing as a person in this conception as relating to their practice, to concepts of creativity and beyond the practice into the student lifeworld.

An analysis of these conceptions in terms of their structural and referential components is shown in Table 6.1. This demonstrates the way the categories have a logical ordering within the outcome space. The community of practice dimension is present in conceptions C, D and E.

Table 6.1   Conceptions of teaching creative practices: outcome space

The structural and referential aspects of the categories of conceptions				
Structural	**Referential**			
Focus of the teaching	Skills	Critical language	Conceptual	Student lifeworld
Giving information to individual students	A			
Developing students through groups (and individuals)		B	C	
Changing students through groups (and individuals)			D	E

Quantitative Method

The additional nine items used to capture aspects of skills and communities of practice were designed using data from interviews with teachers (Drew and Williams, 2002). The interviews were conducted with 44 practice-based teachers of art, design and communication to explore their experience of teaching in terms of both conceptions and approaches to teaching. The interview data on approaches informed this version of the inventory (Art Design and Communication- Approaches to Teaching Inventory – ADC-ATI). Items were added to the ATI using data collected from the interviews which pertained to aspects of teachers' focus on skills and developing communities of practice. The ADC-ATI was distributed to teachers of art, design and communication in UK universities and institutes: 73 returned questionnaires were analysed. The ATI (Prosser and Trigwell, 1999) was developed to measure the variation in the ways teachers approach their teaching in a particular situation.

It has 16 items.

- Eight on a subscale describing an approach which is intended to change student's conceptions or ways of seeing things through a focus on the student. This subscale is labelled conceptual change/student-focused or CCSF for short.

- Eight on a subscale describing an approach concerned with information transmission and a teacher focus, labelled ITTF.

The full 16-item inventory and its use as a relational instrument is discussed by Prosser and Trigwell in their book (1999). The declared aim of this research is to seek relations between these areas, that is, student learning and teaching, in order to derive conclusions for the development of teachers. Research into teacher approaches therefore acknowledges teachers' conceptions and consequently aims to help to improve student learning outcomes by coming to a better understanding of what it is that teachers think and do when they go about their teaching. The only issue for the use of the ATI in the art, design and communication context was the language used in the inventory itself. An adaptation of those items had already been undertaken for a study into approaches to teaching in design contexts (Trigwell, 2002) and as a consequence I gained permission from the inventory's authors to use this version.

I added to this version of the inventory items which pertained to aspects of teachers focus on skills and developing communities of practice. I added items to the ATI using data collected from the interviews, in other words, I used statements made by interview respondents in the approaches section of their interviews and clarified aims, intentions and important issues relating to teaching which focuses on skills or on developing communities of practice.

A factor analysis of the responses to the nine items (Table 6.2) shows that three items which focus on skills development (3, 6 and 9) load heavily on Factor 2, and four items which focus on developing a community of practice (15, 18, 24 and 25) load heavily on Factor 1. These items were combined to form scales, called Skills and Practice respectively.

Table 6.2   **Principal components factor analysis (with varimax rotation) results of the nine skills/practice inventory items**

	Factor		
Item	**1**	**2**	**3**
3	−0.098	**0.810**	−0.029
6	−0.040	**0.856**	−0.036
9	0.126	**0.681**	0.337
12	0.162	0.137	**0.740**
15	**0.738**	0.026	−0.159
18	**0.811**	0.069	0.255
21	0.126	−0.020	**−0.578**
24	**0.574**	0.316	−0.492
25	**0.798**	−0.273	−0.015

N = 73; Eigenvalues >1.00.

I added four items relating to practical skills development (items 3, 6, 9 and 12). Following analysis item 12 was not used in the Skills Scale.

Skills Items

3   My aim in this subject is to develop students' technical competence in basic skills.

6   I think that an important reason for running teaching sessions in this subject is to demonstrate technical procedures correctly.

9   Being able to use the basic skills is a key aim I have for students in this subject.

12   In this subject, I think it is important for students to have opportunities to practise their skills with my support.

I also included five items relating to developing communities of practice (items 15, 18, 21, 24 and 25). Following analysis item 21 was not included in the Practice Scale. Item 25 helped respondents to discern whether a *Practice* approach was more or less important to them than a *Skills* approach.

Practice Items

15   I feel that it is important for students to experience the practice in a 'real-world' situation in this subject.

18   In this subject I help students apply their skills to 'real-world' projects.

21   I think it is important in this subject for students to develop their practice through individually negotiated study.

24   Getting students to think and act like a practitioner is my aim in this subject.

25   In this subject I feel it is more important for students to engage with 'real-world' projects and to act like a practitioner than to develop and practice basic skills.

Results

The structure of the scales has been explored by factor analysis and correlation analysis. In this section, overall scale scores for all four scales are further subject to Principal Component Analysis PCA factor analysis (Table 6.3). The four scales are also analysed in a correlation matrix (Table 6.4).

Table 6.3   **Factor analysis of all four scales**

	Factor	
Scale	**1**	**2**
CCSF	−0.249	**0.786**
ITTF	**0.872**	0.168
Practice	0.244	**0.796**
Skills	**0.866**	−0.171

Extraction method: principal component analysis. rotation method: varimax with Kaiser normalization, a rotation converged in three iterations. Two factor solution: Eigenvalues >0.1.

The figures shown in bold are for loadings above 0.5. Information transmission/teacher-focused approach to teaching and Skills load heavily on one factor and a conceptual change/student-focused approach and Practice load heavily on the other factor. This factor analysis says nothing about the individual teachers, but does say that more of a student-focused approach is related to more time spent on real-world and practitioner-related problems. Conversely a teacher-focused approach is related to the development of skills.

This result has implications for the development of teachers of practice-based subjects if both a student-focused approach to teaching and a practice orientation are observed to have strong relations. This set of relations in the factor structure of the scales is further explored in the correlation matrix of the scales (Table 6.4).

Table 6.4   **Correlation analyses (Pearson, r) for the four scales in the ADC-ATI**

	Scale			
Scale	CCSF	ITTF	Skills	Practice
CCSF	–	–0.05	–0.22	0.27*
ITTF		–	0.56‡	0.23
SKILLS			–	0.02

N = 73; *p <0.05, †p <0.01, ‡ p <0.001.

CCSF approach scores correlate positively (r = 0.27) and statistically significantly (p <0.05) with a focus on using real-world problems (*Practice*). CCSF approaches also correlate positively (r = 0.36) and statistically significantly (p <0.01) with the view that learning to act like a practitioner and to tackle real-world problems is more important than the development of skills (i25) (not shown in Table 6.4). However, these teachers still do develop and practice basic skills with their students, but in the context of the studio- or project-based learning. This finding helps us to understand that in the teaching of creative practices, the student-focused approach also aligns with an approach in which teachers encourage their students to learn through authentic practices (real-world projects).

Discussion

The conceptions of teaching which are described as demonstrating the community of practice dimension illustrate how teachers see learning as engaging with a practice, by exemplars, stories, narratives and through experience. The process of learning becomes one of apprenticeship to the practice, by engaging with the real-world practice and understanding the process through narration, collaboration and social construction (Billett, 2001; Lave and Wenger, 1991).

There is a widely held view in university level teaching that a student-focused or student-centred approach helps students to develop as individuals and also fosters approaches to learning which can lead to higher quality learning outcomes (Trigwell, Prosser and Waterhouse, 1999). From the study described here, it can also be added that in the teaching of creative practices the student-focused approach also aligns with an approach in which teachers encourage their students to learn through authentic practices (real-world projects). If teachers of these subjects value the induction of their students into the community of their practice then it also follows that they should develop both a student-focused approach and a related practice focus to their teaching. This has significance for the development of teachers in these subjects if high-level student learning outcomes in practice-based courses are a desired aim for the teaching.

When teachers describe their approach as being more student-focused, they spend more of their teaching time on real-world and practitioner-related problems. On the other hand, when teachers describe their approach as being more teacher-focused, they report adopting a focus mainly on skills development. It should be emphasised here that most or all of the teachers in this study do develop skills with their students, but those with a student-focused approach focus more on inducting students into the community of practice by using real-world projects and studio- or practice-based approaches. If teachers of these subjects value the induction of their students into the community of their practice then it also follows that they should develop a student-focused approach and a related practice focus to their teaching.

This study confirms the views held by both Wenger (1998) and Billett (2001) that a skills-based approach to learning to practice is simply not enough on its own. There is also evidence here that a skills-based approach corresponds with an Information Transmission/Teacher-focused approach to teaching. Those teachers who do integrate skills into real-world projects and studio- or practice-based approaches, help learners develop competence in those skills so that they can construct an experience of meaning.

The approaches to teaching scores in art, design and communication obtained in this study, as with one previous study (Trigwell, 2002), show high levels of adoption of student-focused approaches. These teachers are describing their approaches in terms of using time to question students ideas, of using difficult or undefined examples to provoke debate, of engaging in discussions with students, and of assessing students in ways that get at their changing conceptual understandings.

In all reported cases of its use, including the results from this study, described above, the Approaches to Teaching Inventory yields interpretable data in the form expected using the educational principles from which it has been developed. For example, Conceptual Change/Student-focused approaches are found to relate positively with students' deep approaches to learning (Trigwell et al., 1999), with perceptions of a manageable workload, some control over what is being taught, a manageable class size and small variation in student characteristics (Prosser and Trigwell, 1997) and with teacher learning (Trigwell, 2002). From this study can also be added the correlation with a focus on the development of professional knowledge for the real world.

References

Billett, S. (1998) Situation, social systems and learning, *Journal of Education and Work*, 11, 255–274.

Billett, S. (2001) Knowing in practice: re-conceptualising vocational expertise, *Learning and Instruction*, 11, 431–452.

Drew, L. (2004) The experience of teaching creative practices: conceptions and approaches to teaching in the community of practice dimensions, in A. Davies (ed.), *Enhancing Curricula: Towards the Scholarship of Teaching in Art, Design and Communication in Higher Education*, pp. 106–123. London: Centre for Learning and Teaching (CLTAD).

Drew, L. (2000a) Do development interventions shape conceptions of teaching in art and design?, in C. Rust (ed.), *Improving Student Learning: Improving Student Learning through the Disciplines*, pp. 230–243. Oxford: Oxford Brookes University, Oxford Centre for Staff and Learning Development.

Drew, L. (2000b) A disciplined approach: learning to practice as design teachers in the university, in C. Swann and E. Young (eds), *Reinventing Design Education in the University*, pp. 187–193. Perth, Western Australia: Curtin University of Technology.

Drew, L. and Williams, C. (2002) Variation in the experience of teaching creative practices: the community of practice dimension. Paper presented at 10th Improving Student Learning Symposium: Improving Student Learning: Theory and Practice – 10 years on, Brussels, Belgium, 5–7 September.

Lave, J. (1993) The practice of learning, in S. Chaiklin and J. Lave (eds), *Understanding Practice*, pp. 3–32. Cambridge: Cambridge University Press.

Lave, J. (1996) Teaching, as learning, in practice, *Mind, Culture and Activity*, 3, 149–164.

Lave, J. and Wenger, E. (1991) *Situated Learning: Legitimate Peripheral Participation*. Cambridge: Cambridge University Press.

Marton, F. and Booth, S. (1997) *Learning and Awareness*. Mahwah, NJ: Lawrence Erlbaum.

Prosser, M. and Trigwell, K. (1993). Development of an approaches to teaching questionnaire, *Research and Development in Higher Education*, 15, 468–473.

Prosser, M. and Trigwell, K. (1997) Perceptions of the teaching environment and its relationship to approaches to teaching, *British Journal of Educational Psychology*, 67, 25–35.

Prosser, M. and Trigwell, K. (1999) *Understanding Learning and Teaching: The Experience in Higher Education*. Buckingham: SRHE and Open University Press.

Reid, A. (2000) Musicians' experience of the musical world: relations with teaching and learning, in C. Rust (ed.), *Improving Student Learning: Improving Student Learning Through the Disciplines*, pp. 179–185. Oxford: Oxford Brookes University, Oxford Centre for Staff and Learning Development.

Reid, A. and Davies, A. (2000) Uncovering problematics in design education: Learning and the design entity, in C. Swann and E. Young (eds), *Reinventing*

Design Education in the University, pp. 179–185. Perth, Western Australia: Curtin University of Technology.

Rogoff, B. (1990) *Apprenticeship in Thinking – Cognitive Development in Social Context*. New York: Oxford University Press.

Trigwell, K. (2000) A phenomenographic interview on phenomenography, in J. Bowden and E. Walsh (eds) *Phenomenography*, pp. 47–61. Melbourne: RMIT Publishing.

Trigwell, K. (2002) Approaches to teaching design subjects: a quantitative analysis, *Art, Design and Communication in Higher Education*, 1, 69–80.

Trigwell, K. and Prosser, M. (1996). Changing approaches to teaching: a relational perspective, *Studies in Higher Education*, 21, 275–284.

Trigwell, K., Prosser, M. and Taylor, P. (1994) Qualitative differences in approaches to teaching first year university science, *Higher Education*, 27, 75–84.

Trigwell, K., Prosser, M. and Waterhouse, F. (1999) Relations between teachers' approaches to teaching and students' approaches to learning, *Higher Education*, 37, 57–70.

Shreeve, A., Wareing, S. and Drew, L. (2008) Key aspects of teaching and learning in the visual arts, in S. Fry, S. Ketteridge and S. Marshall (eds), *A Handbook for Teaching and Learning in Higher Education: Enhancing Academic Practice*, pp. 345–362. London: Routledge.

Wenger, E. (1998). *Communities of Practice: Learning, Meaning and Identity*. Cambridge: Cambridge University Press.

Chapter 7

Transformative Practice as a Learning Approach for Industrial Designers

KAREN BULL

Design Learning in a Contemporary Context

Industrial design teaching and learning is about the transformation of the individual design learner from aspiring designer to being one that is professionally ready. Core to this is a process that requires the integration of both holistic and linear ways of thinking through practice and especially design projects, experiential problem-solving and creative experimentation. This chapter is concerned with how this can be achieved in a contemporary educational setting to create a design curriculum that embraces a transformative approach to design learning. It must be one of co-ordinating the teaching of multiple disciplines, promoting subject interdependency and highly practical modes of learning that are strongly aligned to developing aspirations and professional communities of design practice.

Such a goal requires careful implementation within an UK higher education (HE) framework that predominantly focuses on the efficiency and accountability of a modular teaching, learning and assessment framework. It is constructed around an accumulative credit framework associated with building programmes of study from individual modules that combine into a complete degree course (see Figure 7.1).

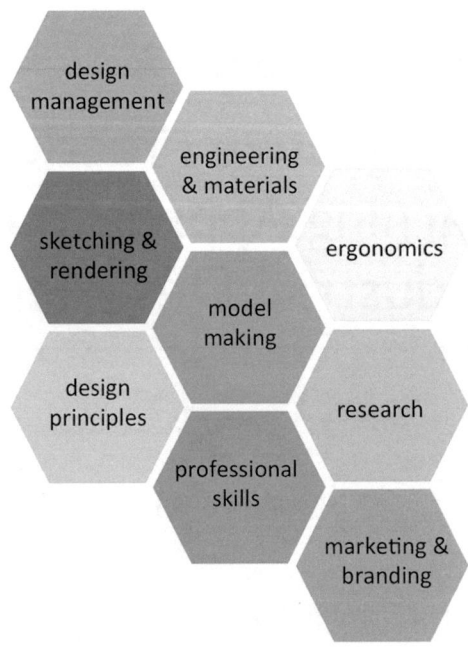

Figure 7.1   A modular learning framework

It is appealing in terms of rationalising the curriculum 'so that a single set of lectures [are taught] in a subject common to students in a variety of disciplines' (Trowler, 1998, p. 91). While modularity offers flexibility in some ways, and the self-contained modules are 'administratively neat' in practical and evaluative terms, this is not like design in practice. The potential fragmentation of the learning process could be seen as a potential danger (Billing, 2007; Hennessy, Hernandez, Kieran, and MacLough, 2010; Roderio, 2008) which could manifest itself in a surface approach to learning and in turn a difficulty relating 'theory to practice' (Roderio, 2008).

Design learning is well known for being flexible and centred around developing the individual learner through practice with a strong student–tutor relationship, as seen in Atelier-style teaching and apprenticeship models. While these approaches may have particular strengths in terms of personalised learning they can be outdated, hard to frame and inefficient in a modern context. It is clear that pressures to teach larger cohorts of students that have high expectations within tight system constraint raises questions about how such a transformative learning experience can be maintained in contemporary educational settings (Hennessy et al., 2010) and this is pertinent to design learning where emphasis needs to be on experiential learning and studio practice to help transform design knowledge into practical skill and capability, and in turn produce confident and professionally ready designers.

Constructive, Abductive and Solution-led Thinking

Fundamentally design teaching needs to provide a platform that can support the teaching of skills and knowledge but also frame the mechanisms for teaching a 'designerly way of thinking'. Cross (2006) suggests that design thinking involves the 'tackling of "ill-defined" problems through a "solution-led" problem solving approach'. This ranges from holistic and explorative approaches centred around the highly 'synthetic' forms of thinking and doing, and the analytical techniques that help to develop, align, determine and structure design activities. The approach is constructive (centred on creating new ideas based on old ideas) and abductive (exploring what could possibly be true) (Curedale, 2013). All of this is centred strongly on an experiential model of learning and especially the need to practice what we know.

As Curedale (2013) demonstrates in Table 7.1, keywords for design thinking are strongly centred around user empathy, exploration, being inquisitive and remaining open-minded. Owen (2007) reinforces this by characterising design thinking in part as holistic and generalist approach which integrates human empathy, adaptivity, inventiveness, contextual appreciation and the ability to visualize and maintain an optimistic predisposition when engaging with tasks. These are hard elements to structure within a constrained modular system that is centred around smaller packets of specialist learning. What is clear is that much of this requires students and professionals alike have a 'toleration of uncertainty' (Osmond and Turner, 2010). As Baeck and Gremmet (2012) say, it is necessary to be 'comfortable when things are unclear or when you do not know the answer'.

Table 7.1   Core attributes of design thinking (Baeck and Gremett, 2012)

Ambiguity	Being comfortable when things are unclear or when you do not know the answer	Design Thinking addresses wicked=ill-defined and tricky problems
Collaborative	Working together across disciplines	People design in interdisciplinary teams
Constructive	Creating new ideas based on old ideas, which can also be the most successful ideas	Design Thinking is a solution-based approach that looks for an improved future result
Curiosity	Being interested in things you do not understand or perceiving things with fresh eyes	Considerable time and effort is spent on clarifying the requirements. A large part of the problem solving activity, then, consists of problem definition and problem shaping
Empathy	Seeing and understanding things from your customers' point of view.	The focus is on user needs (problem context)
Holistic	Looking at the bigger context for the customer	Design Thinking attempts to meet user needs and also drive business success
Iterative	A cyclical process where improvements are made to a solution or idea regardless of the phase.	The Design Thinking process is typically non-sequential and may include feedback loops and cycles (see below)
Non judgmental	Creating ideas with no judgment toward the idea creator or the idea	Particularly in the brainstorming phase, there are no early judgments
Open Mindset	Embracing design thinking as an approach for any problem regardless of industry or scope	The method encourages 'outside the box thinking' ('wild ideas'); it defies the obvious and embraces a more experimental approach.

An important factor added to this perspective on the characteristics of design thinking and education includes the value of design thinking beyond the design of the artifact. The consideration of design as a 'way of thinking for productive purposes' rather than just a way of producing 'things' is significant. Curedale defines design thinking as

> *an approach that seeks practical and innovative solutions to problems. It can be used to develop products, services, experiences and strategy. It is an approach that allows designers to go beyond focusing on improving the appearance of things to provide a framework for solving complex problems. Design thinking combines empathy for people with their context to discover new insights. It drives business value. Companies such as GE, Target, Procter and Gamble IDEO and Intuit have successfully applied this approach to design. (Curedale, 2013)*

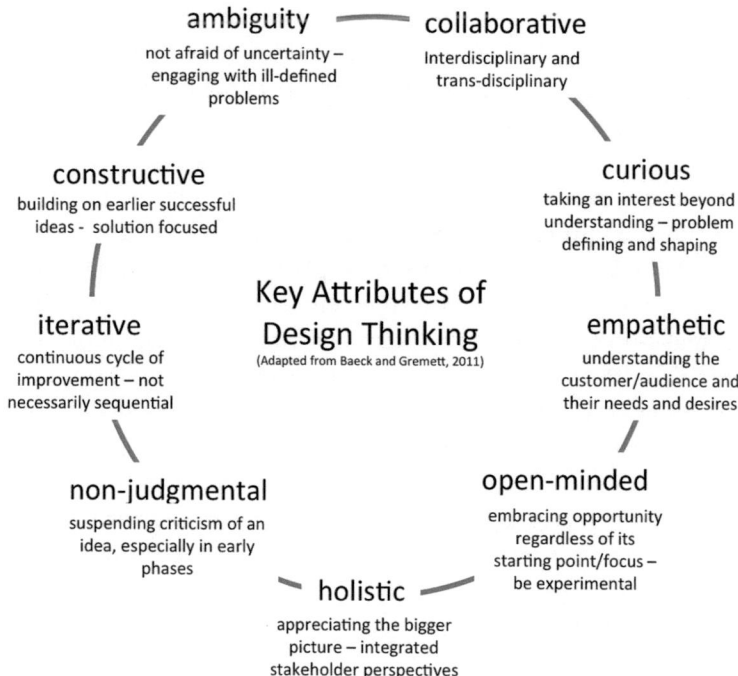

Figure 7.2 **Core attributes of design thinking. Adapted from Baeck and Gremett (2011)**

It is clear that becoming a successful designer depends strongly on individual capability to think in a designerly way (holistic and generalistic) as well as the specialist design skills and knowledge to translate and develop ideas, not a way of thinking that is easy to achieve within the delivery of modular packets of learning. A designer may at times be dealing with unexpected or unfamiliar problem scenarios. This requires an adaptable and confident individual who can draw upon a range of knowledge, capabilities and experiences and combine them together using different modes of thinking to produce creative solutions. A significant emphasis for educators is therefore placed at the level of the individual design student, how they are nurtured to engage with uncertainty and how they develop the personal confidence to push ideas from their own developed knowledge and experience base. This is important in terms of preparation to enter a professional design community. Tovey and Bull (2010) describe this as being about helping students to develop 'a passport to professional practice', and at the heart of this is acquiring a shared understanding and appreciation of common approaches to practice built around a strong personal and professional identity.

This is where the emphasis upon a 'transformative practice-based approach' becomes core to design education and preparing students for entry to their 'professional community of design practice'. A way of thinking about this is to consider in simplistic terms how plants grow (see Figure 7.3).

The plant is dependent on a number of integrated elements, e.g. air, water, light, warmth and nutrients. Omit one and the plant will not grow successfully. The right balance is required to support the plant effectively and it needs a place and time to grow. In the context of the plant, it is the soil that must provide the right make-up to support the plant and provide a regulatory system for delivering elements such as water and nutrients. In terms of design learning there is a similar need for the regulated delivery of skill, knowledge, experience, creativity and criticism within a good growing space. In simplistic terms the soil (*studio and safe space to stretch out ones roots*), the seed and plant (the *student who contains lots of energy and aspiration*), nutrients and water (*teaching, input, inspiration*), sun and warmth (*professional outlook and focus*), time (*time to develop and grow without limitation*). With all these elements in place you have a culture for the transformation of an aspiring designer to one that is ready to engage in their profession.

Figure 7.3   The elements for growth

Fundamentally this foundation for knowledge acquisition and transformation could be considered in terms of action. In the context of the plant analogy this is the chlorophyll that forms the chemical change in the plant, producing the energy for growth. In this discussion 'designerly thinking' is the active ingredient which underpins the transformation and here it is articulated as a dual-process model of cognition (Tovey, 1984) that draws together the analytical and synthetic through a dialogue, journey or space where creative ideas are generated. At the heart of this is the 'solutioning mentality' that designers are well known for. It is not easy to articulate the approach in scientific terms and it is not easy to measure the creative process, but observation indicates that it is centred on applying knowledge and experience through a dialogue centred on creative praxis. This is why a transformative practice approach to design learning should be adopted to provide the 'space' for students to iteratively develop their design capabilities and skills through design practice and thus gain the confidence to practically and creatively deal with increasingly sophisticated or even troublesome design issues, whether they be user, visual, practical or technically focused.

Building on the Baseline Capability of a Design Student

Design students often arrive at university with a baseline capability to design, e.g. they can visualise in 2D and 3D, think imaginatively and develop concepts based upon simple investigative activities. It is from this baseline that tutors have to help students recognise their position, skills and capabilities relative to settings such as industry, business, engineering, technology, manufacturing, users, markets, trends, aesthetic and brand considerations. Much of the work is about 'opening up' student awareness of their operating context. Many students arrive with a fixed aspirational view, e.g. I want to become a studio designer for Aston Martin. The true nature of design student learning is actually far more complex. The setting or 'soil' for designers must be supportive and about broadening design awareness and engagement with design processes. This has to be achieved through a mix of knowledge acquisition, skills and capabilities development and practice, critical evaluation and development of an experience base that provides the foundation for the so called 'intuitive' decision-making capabilities that designers recognise as characteristic of their practice (Cross, 2006).

It is this experience base combined with highly personal perspectives, viewpoints and approaches to design that creates the design personality that is often talked about in the industrial design field. A significant part of design educational practice is about allowing students to develop as individuals with unique portfolio assets and the preparedness to engage with industry that makes them attractive to employers. For obvious reasons employers seek fresh vision and imagination as instrumental to building on innovation and competitiveness within their own business, but they also need to know that students are professionally aware in a global context.

Design Thinking for Global Competitiveness

As a result design education today has to be, in part, about preparing students to thrive in a setting of global economic competitiveness. At the heart of this is a fusion of critical thinking and problem solving, communication, and creativity and innovation (P21, 2013). A study by the OECD suggests that design is recognised as a knowledge-based capital that can be used to drive innovation and growth, and that it is integral to all stages of the business process from research to manufacture and after-sales services. It also recognized that in the UK for example, design spending might be more than twice as large as business spending on R&D and it plays an important role in innovation and performance (OECD, 2012). The same study indicates that almost half the businesses in the UK believe that design helps to increase market share and turnover. There is global recognition of design in terms of strategic economic importance. Therefore the breadth of roles associated with industrial design also needs to be recognized and it is important to see beyond the traditional role of the industrial designer as developer of the physical artefact to one that may influence creative direction, strategic business, creative think-tanks across a range of commercial and organizational contexts, as well as informing experience-centred products and services. As a result it is important to see that industrial design education may take students beyond practice to utilizing creative ability in other more strategic directions. This demonstrates the role of design educators to not just turn out good designers but to ensure students are highly capable of recognizing their own design thinking capabilities in the wider commercial environment. This further supports the need for a transformative approach. The student needs to become aware of the opportunity to build upon their ability to generate ideas at the physical level and embrace the challenge to offer creative insight at a more strategic level.

A Good Setting for Industrial Design Learning

A good setting for industrial design teaching is already well established. There are variants in terms of emphasis (e.g more technical- or engineering-focused, arts or human-factors emphasis), and many programmes follow the 'project' as the space for learning. Rarely however do we stop to reflect on the complex nature of design learning and what brings together the wide-ranging elements of learning and practice within a modular setting (see Figure 7.4).

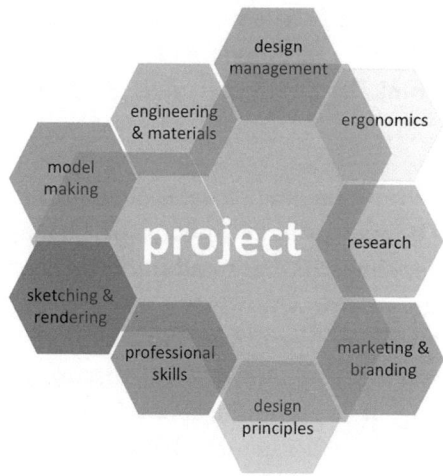

Figure 7.4   Modules brought together by the project

Coventry School of Art and Design has had an Industrial Design department serving undergraduates for forty years and also has strong postgraduate provision. The department covers transport, product and interior design and is world renowned with regard to automotive design. The courses are strongly centred on the project and a very balanced emphasis of study that recognizes the need to provide good learning spaces that underpin the elements of design thinking in a transformative learning context. The Product Design course especially aims to balance, through critical and creative thinking, the visual and technical against a strong user experience or requirements focus.

The department recognizes its breadth and prides itself on a wide range of specialist expertise from designers, engineers and ergonomics to CAD and form specialists. The department was recognized by HEFCE (Higher Education Funding Council for England) for excellence in teaching and learning in 2005 and at that point it established CEPAD, the Centre of Excellence for Product and Automotive Design. The research and evaluation conducted by CEPAD helped the department to reflect upon the curriculum and refine it to align more strongly with the transformative and the 'dual-processing' approach described. Within that structure the transformative practice approach to learning was first articulated and appropriate significant innovations and facilities were recognized and mapped within the curriculum. For example: the use of gateways for more holistic assessment and feedback; the use of audio and peer-centred feedback models for more personalized learning; and the increasing awareness of the MDes Fourth Level (enhanced by inclusion of some level 7 modules) of study as a means for students to focus upon wider and more critical and creative questioning of design problems and innovation starting points, e.g. 'Think Big' philosophy as highlighted by Tim Brown (CEO of IDEO).

Industrial design education at HE level is often non-linear, highly tacit and practitioner-focused. It involves large amounts of what appears to be intuitive or non-measurable decision-making. In addition the setting for this teaching is normally the design studio, which is not just a room but an environment equipped for conceiving and resolving ideas most often in a collaborative manner. Shreeve, Wareing and Drew (2008) discuss the studio setting for design learners as being, at its optimum, about developing engaging discourse and recognizing the tutor as a facilitator rather than merely 'dropping pearls of wisdom'. This apparently informal setting can be an entirely new experience for design learners entering HE especially from international contexts. At the heart of the studio environment is the sense that it is okay to take risk and that one can play with ideas and test out one's capabilities to design and think about design. A danger is that within the UK university teachers often have a high expectation based on our own educational habitus that everyone has encountered the studio-culture. Experience tells us that some students, mainly those who join us from international contexts where there is a higher level of rote learning or a larger power-distance relationship between the lecturer and the student, find adapting to such a learning environment more challenging (Bull and Osmond, 2013). At Coventry we are now exploring educational models that will allow international students who are new to such practice to engage with studio learning, critical and creative risk-taking before embarking on the formal accountable aspects of their courses.

This unfamiliar territory for learning is made more challenging by the complex nature of design activity itself. As Buchanan described (1992), central to designing are 'wicked problems' which are at their foundation 'ill-formulated', difficult to separate out and confusing. Students during prior learning have often dealt with fairly well-defined problem starting points in the context of design solutioning but at HE level quickly get exposed to more challenging and difficult problem scenarios – in some instances the process involves a significant phase of 'problem definition' where limits and contexts of problems are unclear without huge effort to comprehend them. For example, many students from China who have prior experience of design learning are more familiar with being given very short (sometimes day-long) projects that are mainly centred around skill development, e.g. drawing, drafting, CAD or visualization. UK HE students are often familiar with the significant stages of the design process from investigation to resolution to ideas but have not had the freedom to challenge a brief or engage with a deeper level of critical design questioning.

Such discomfort with learning is often evidenced in new HE design learners by a lack of design confidence brought about by uncertainty during the conceptual and incubation phases of the design process (see Figure 7.5).

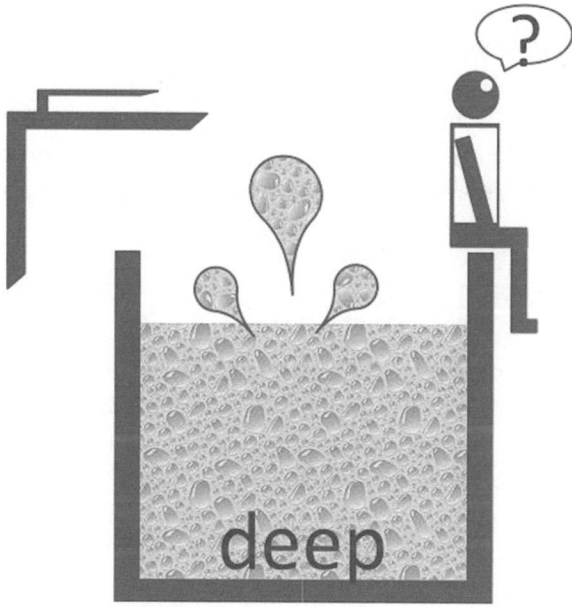

Figure 7.5   Stuck in a liminal space

Osmond defines this as being 'stuck in the bubble' (Osmond and Turner, 2010) a phase involving the uncertainty of searching for solutions and alternatives and frames this as a significant threshold concept within industrial design education 'The Toleration of Design Uncertainty' as defined during the CEPAD evaluation (Osmond and Turner, 2010). As a result it makes sense to support design learners by providing for them plenty of space for experimentation, reflection, trial and error built around models of good design discourse and problem-centred exploration. Provision of 'safe spaces' for creative learning that are constructed around positive and constructive peer-, tutor- and practitioner-based discourse and practical project engagement can enable students to apply creative reasoning and conceptual development without fear of failure or humiliation.

Learning to 'Unthink' Ways of Doing Design Thinking

This chapter has proposed that the confidence to design can be gained through an experiential model of transformative practice which provides spaces for both physical and cognitive creative risk-taking, design play and solutioning. Transformative learning itself is about learning to develop autonomous thinking and the process of effecting change in a 'frame of reference' (Mezirow, 1997). In the context of design practice this can be defined as learning to 'unthink' ways of 'doing design thinking' as previously taught. In HE we inculcate a more challenging exploration of design alternatives with emphasis on trial and error, design-risk and incubation phases of design activity resulting in more space for reflection, intensive exploration and questioning of alternatives as well as the application of a wide range of creative solutioning techniques.

Building a Scaffolded Approach

Recognition that students have to transform from a position of 'aspiring to be a designer' and recognize their own skills and capabilities within in the broad territory of industrial design to becoming a fully confident members of their professional community of practice is core to a good design curriculum. To achieve this good integrative 'scaffolds' need to be put into place to allow students to build upon their knowledge and capabilities in a holistic and meaningful way. Such an approach will help students to develop over time, certainty to exhibit design solutions within the professional community and result in the acquisition of their 'passport to professional practice' (Tovey and Bull, 2010), will show them to be confident, creative and above all prepared to publish in the context of their community of practice.

The Gateway Assessment model of learning is a scaffold example. This is a technique applied in the context of Coventry University Industrial Design curriculum to support formative or summative assessment. Such Gateways may take place two or three times a year. During these events students exhibit all elements of their learning for that period. Tutors work as a collaborative team to conduct the review of work either with or without the students present. The goal is to look at the work in a holistic manner and to evaluate the progress of the student against a wider set of learning objectives. It becomes a 'portfolio review' in practice. The idea is that the tutors can more holistically identify strengths and areas for improvement for each student and through either written, verbal or audio forms of feedback provide clearer frames of reference for the student to develop their own aspirations and professional foci. At Level 1 of study this might be more generic, helping a student to develop specialist pathways. In the final year this might be to nurture a specific and individual range of skills or capabilities foci within a student's final major project. The Gateway model is also a way of working more collaboratively utilizing peer assessment and peer support during assessment to develop both student awareness of learning requirements or to support students in capturing key advice while leaving tutors free to provide richer and more in-depth guidance without having to focus on writing up feedback notes as they speak (Osmond and Clough, 2012).

A Designerly Way of Knowing and Behaving

The reason the challenge of uncertainty may exist (it is similar to the feeling of 'designers block' or being afraid put the first mark on a blank sheet of paper within new HE design learners) is because solutioning is by nature a troublesome activity integrating different modes of thinking. As introduced, Tovey's model (1984) is defined as a dual-processing model which focuses on drawing together the analysis–synthesis aspects of design thinking and activity (Tovey and Bull, 2010). At its centre is the 'Uncertainty Threshold'. It assumes that the two halves of the brain both work to solve design problems. Each uses a different cognitive mode of processing, the left being concerned with the verbal, analytic, linear, serial and the right being about synthesis and the manipulospatial, synthetic, holistic, parallel and diffuse. Tovey explains that it is the apositional matching of the requirements between the two halves that allow designers to engage with 'wicked problems', thus delivering a balance between gathering data, optimizing evaluations and solutioning activities. This 'matching process' is where the lack of confidence to deal with design uncertainty can exist. For the purposes of exploring the design learning experience it is helpful to build upon Tovey's model with an analogy, that of following a recipe to produce a sophisticated dish (see Figure 7.6). A recipe is constructed of ingredients of various types, forms and quantities and the other is method. A series of actions are completed in an effective combination to produce the 'dish' in question.

Putting the analogy into practice, the ingredients provide the substance of the creative activity. They are centred on factual, analytic, logical, evaluative activities. They are used to define, articulate and provide a framework for the underlying parameters, structure or substance while forming a design solution. In terms of actions, these are the 'synthetic' methods that are applied to build upon, challenge, explore, reframe, and experiment or push boundaries with that substance. As a beginner the dish prepared may not turn out as expected and it will be a matter of practice, trial and error and making adjustments according to personal preference that will produce the desired result – this is an iterative and solution-focused process centred around the synthetic or holistic modes of thinking. As the certainty and confidence grows the creator will become more adventurous in their goals because they are less afraid to challenge boundaries in their practice, turning their work into culinary arts and challenging existing gustatory effects that might be culturally or contextually specific or even transcend such boundaries.

The design learner is not unlike the novice chef who will follow instructions carefully but without practice may miss important requirements concerning ingredients, or not have the experience to carry out the cooking tasks effectively (e.g. the stereotypical soufflé that doesn't rise – chef Beard says *The only thing that will make a soufflé fall is if it knows you are afraid of it*). In contrast, an experienced chef will know exactly how to match ingredients with actions, challenge the normal ingredients to create new flavours, or apply a different approach to combining ingredients. This is learnt through repeated practice, skill development, individual experience and trial and error. This is not something that can be achieved instantly, it takes time to master and the confidence to take risks. What becomes evident is that the space between the analytical and synthetic becomes highly significant as the transformative space (see Figure 7.5) because the analogy demonstrates the requirement for practical experience and experimentation, safe spaces for creative risk-taking and an effective environment in which to practically draw together these two domains of design thinking to their best advantage. It highlights the need for iteration and repeat experience with increasing awareness of personal achievement to enable the chef/designer to accomplish skill and professional confidence.

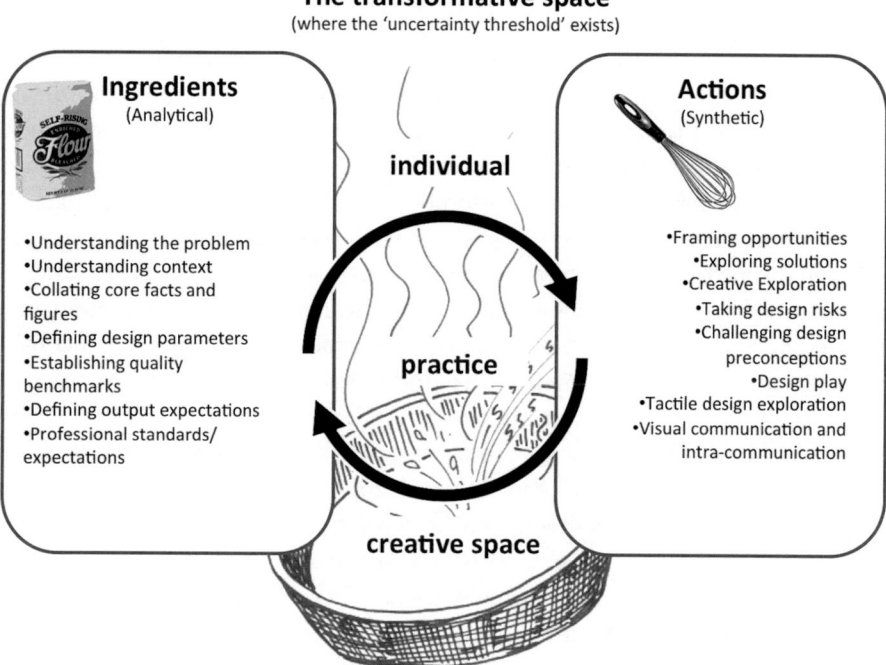

The transformative space
(where the 'uncertainty threshold' exists)

Ingredients
(Analytical)

individual

•Understanding the problem
•Understanding context
•Collating core facts and figures
•Defining design parameters
•Establishing quality benchmarks
•Defining output expectations
•Professional standards/ expectations

Actions
(Synthetic)

•Framing opportunities
•Exploring solutions
•Creative Exploration
•Taking design risks
•Challenging design preconceptions
•Design play
•Tactile design exploration
•Visual communication and intra-communication

practice

creative space

Figure 7.6   The transformative space

Transformative Practice Approaches Within the Industrial Design Curriculum

As mentioned the Industrial Design Department at Coventry University has recognized this transformative practice based approach to learning for a number of years and it is embedded within our highly valued studio-culture approach to learning (Tovey and Owen, 2006). The studio as a creative space is at the core. The environment is carefully constructed around flexible workspaces that enable group collaboration, individual working or seminar activities. Pin-up spaces and portable display screens allow the spaces to be used according to purpose. Professionally furnished break-out rooms allow students to work professionally as groups and emulate an industry experience. In terms of curriculum, the structure of Gateways as previously described enables a holistic teaching, learning and assessment approach, allowing students the freedom to engage in increasingly more complex and real-world/industry-focused design problems.

A scaffolded approach takes students from Level 1 where study begins by embedding a studio culture and reinforcing a strong design skills base and simple creative tasks. Design Roots (see Figure 7.7) is an example of an activity at that level which focuses almost entirely on the ability to transform design knowledge into a practical studio-based learning context. This two-week non-assessed assignment was introduced to Level 1 students in order to introduce them to the historical context of automotive and transport design along with an introduction to significant key skills such as writing and investigation. Previous approaches which were centred on the chronological teaching of design proved challenging as students did not translate their taught knowledge into their design practice. This transformative practice-based approach aims to bring design history teaching into the studio environment. It achieves this by embedding a practical design brief that requires the interpretation of design history materials through a collaborative process of negotiation investigation and designing (Johnson, Bull and Osmond, 2013). The result was an exhibition event in the studio environment that became rich material for future design project inspiration (see Figure 7.7). It was noted that students engaged more deeply with their theoretical knowledge and even formed their own design principles for translation into design practice. The key contributor to the student engagement with design history learning was the collaborative, dialogue-centred and solution-based activities they participated in. The team approach reduced anxiety and enhanced their confidence to generate dynamic and original sculptures at an unfamiliar scale without any preconceived idea of a solution at the project introduction.

Figure 7.7   Design roots sculptures

In addition reflective approaches to recognizing personal aspiration as part of that transformative approach to learning are introduced at Level 1, e.g. reflective and self-diagnostic exercises also help to focus on holistic awareness of skills in line with professional aspirations. For example the Wheel of Design tool (Bull, Barrett and Osmond, 2013). This group-based discussion and reflection tool (see Figure 7.8) aimed to help students diagnose their own pathways for learning by getting them to reflect upon their personal skills and capabilities in a holistic manner in relation to their aspirations. Students were encouraged to discuss their aspirations in a peer-based environment to define pathways for development and in turn personal action plans. It also formed the basis of forming peer-based support groups. All of these examples are at the centre of a transformative practice-based approach, helping student's to develop identity and to build their confidence through an appreciation of their future alignment to a professional community of design practice.

Level 2 focuses on professional awareness through live projects and collaborative and peer-based learning. Assessment Buddies (Osmond and Clough, 2012) is an example of an educational approach which loosely emulates collaborative industry practice, helping bring together the design learning experience. Internationally focused collaborative design projects across international institutions have been incorporated to promote transcultural and transdisciplinary perspectives on design practice, such as a Level 2 design project with EAFIT University in Colombia where students formed teams to engage in a collaborative design project to design a piece of transport (Atkinson et al., 2010).

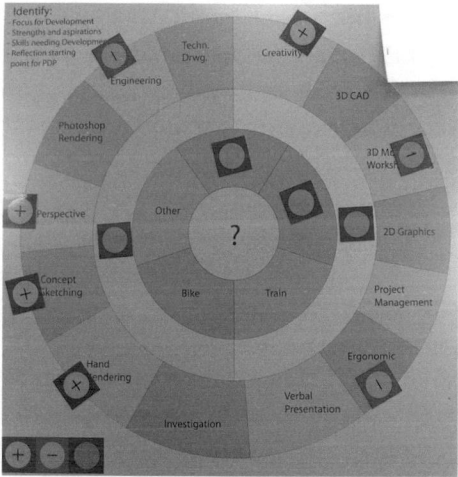

Figure 7.8   The wheel of design

Level 3 (of four) focuses on professional experience and engagement, encouraging live projects and competitive professional placements (a flexible structure supporting twenty weeks, six months or even year-long placements) or international exchange through schemes like Erasmus. This serves to increase student's outward-facing perspective on design aspirations and career goals.

At Level 4 (BA and MDes) the year is strongly centred on individual practice and portfolio production. This is the time where students are strongly encouraged to form a clear personal strategy and affinity with their professional community of practice. It is felt that by this stage students are strongly confident in terms of their creative practice and able to embed simultaneously the holistic and linear modes of design thinking. This is especially noted within our Level 4 MDes cohort who focus their ability around the deeper application of both cognitive styles, linear and simultaneous, through problem definition and solutioning exercises which are underpinned by fairly in-depth design thinking techniques. They utilize a series of iterative techniques for asking 'big' questions, exploring research territories and visualizing 'design hunches' to progress towards the development of evidenced design briefs and specifications. They are able to search complex problem spaces and identify opportunity pathways.

Throughout the course there is regular interaction with Coventry's Industrial Design Alumni. A series of high-profile visiting lecturers are invited in to give weekly department wide talks from an industry and professional experience perspective and these are promoted to all disciplines within the department. A strong profile of industry-experienced teaching staff make up the teaching team and where possible work collaboratively. At times they 'co-teach' to purposefully integrate their discipline knowledge. For example, final major project seminars are frequently led by a design and engineering or a design and ergonomics tutor working to draw together students ideas and foci.

Conclusion

This chapter aims to reinforce the benefits of a transformative practice-based approach to learning. The research has been formalized through the University's Centre of Excellence in Product and Automotive Design (CEPAD) and is based on many years of observation as well as a formal longitudinal study of student experience and other related case-study activity, some of which are cited in this chapter. The chapter discussed how this approach will help students to engage with the 'the threshold of uncertainty' which is often connected to the troublesome experiences that are encountered when engaging with design problems and solutioning by embedding a studio-based culture of learning that engages students through practice and experiential learning. This, as has been demonstrated by this chapter, draws together the systematic and the synthetic modes of thinking as defined by the 'dual processing model of design learning' (Tovey and Bull, 2010). This learning environment, which overcomes the fragmentation that can occur on modular curriculum structures, provides most significantly 'safe spaces' for problem-based learning that allow students to take creative risk without fear of failure. It focuses on scaffolding students' learning from aspiration to a strong personal alignment with a professional community of design practice. This is achieved through a Gateway Assessment Model of assessment and feedback embedded into the curriculum that incorporates collaborative and peer-based learning. This helps students and tutors to define learning pathways. As well as experiential learning, global industry awareness is also treated as part of that transformative approach with a variety of industry inputs to the curriculum. This ranges from strong industry experience of staff to placement activities and live collaborative project experience. These elements all serve to reinforce student aspiration, confidence and professional awareness through a transformative practice-based model of learning.

References

Atkinson, P., Osmond, J. and Sierra Zuluaga, L. (2010) Deepening student engagement in a web-based collaborative group transport design project. *Fourth International Technology, Education and Development Conference, 8–10 March 2010*, Valencia, Spain.

Baeck, P. and Gremett, A. (2011) in Degen and Xiaowei Yuan (eds), *UX Best Practices – How to Achieve More Impact with User Experience*. McGraw-Hill Osborne Media, New York, USA.

Billing, D. (2007) Review of modular implementation in a university, *Higher Education Quarterly*, 50(1), 1–21.

Buchanan, R. (1992) Wicked problems in design thinking, *Design Issues*, 8(2), 5–21.

Bull, K. and Osmond, J. (2013). Design education and non-EU students – shifts in teaching practice. DRS// CUMULUS 2013, Second International Conference for Design Education Researcher, 14 to 17 May 2013, Oslo.

Bull, K., Barrett, A. and Osmond, J. (2013) Wheel of design – reflective alignment of design skills with aspirations. International Conference on Engineering and Product Design Education, 5 and 6 December 2013, Dublin.

Cross, N. (2006) *Designerly Ways of Knowing*. London: Springer Verlag.

Curedale, R. (2013) *Design Thinking: Process and Methods Manual*. Topanga. CA.: Design Community College Inc.

Hennessy, E., Hernandez, R., Kieran, P. and MacLough, H. (2010) Teaching and learning across disciplines: student and staff experiences in a newly modularised system, *Teaching in Higher Education*, 15(6), 675–689.

Johnson, C., Bull, K. and Osmond, J. (2013) Cooperative design and communities of practice. *Tenth International Conference of Cooperative Design, Visualisation and Engineering*, pp. 141–152. Springer Verlag. Berlin Heidelberg.

Mezirow, J. (1997) This chapter summarizes the transformation theory of adult learning, explains the relationship of transformative learning to autonomous, responsible thinking (viewed as the central goal of adult education), and discusses practical implications for educators. Transformative Learning: Theory to Practice. *New Directions For Adult and Continuation Education*. 1997(74), 5–12,

Micklethwaite, P. (2005) Discussing art and design educations: themes from interviews with UK design stakeholders, *Journal of Design Education*, 88(24.1), 84–92 .

OECD. (2012) *New Sources of Growth: Knowledge-based Capital Driving Investment and Productivity in the 21st Century*. OECD. http://www.oecd.org/about/

Osmond, J. and Clough, B. (2012) Involving assessment buddies in the assessment of design project work, *Design and Technology Education: An International Journal*, 17(2), 62–67.

Osmond, J. and Turner, A. (2010) The threshold concept journey in design: from identification to application, in J.H. Baillie (ed.), *Threshold Concepts and Transformational Learning*, pp. 347–364. Rotterdam: Sense Publishers.

Owen, C. (2007) Design thinking: notes on its nature and use, *Design Research Quarterly*, 2(1), 16–27.

P21. (2013) *Partnership for 21st Century Skills*, 20 October. Retrieved 21 October 2013 from www.p21.org.

Roderio, C.N. (2008) *Effects of Modularization*. http://www.cambridgeassessment. org.uk/ca/digitalAssets/186732_CVR_RN_Effects_of_modularisation_-_ Final_Report.pdf.

Shreeve, A., Wareing, S. and Drew, L. (2008) Key aspects of teaching and learning in the visual arts, in H.K. Fry (ed.), *A Handbook for Learning and Teaching in Higher Education*, 3rd edn. London: Kogan Page.

Tovey, M. (1984) Designing with both halves of the brain, *Design Studies*, 5(4), 219–228.

Tovey, M. and Bull, K. (2010) Design education as a passport to professional practice. Trondheim: International Conference on Engineering and Product Design Education.

Tovey, M. and Owen, J. (2006) Entering the community of practice of automotive design. *Sixth International Symposium on Tools and Methods of Competitive Engineering*. Ljubljana: University of Ljubljana. Delft University of Technology, Netherlands.

Trowler, P. (1998) What managerialists forget: higher education credit frameworks and managerialist ideology, *International Studies in Sociology of Education*, 8(1), 91–110.

Chapter 8

Industrial Design and Liminal Spaces

JANE OSMOND

Coventry University

I had a sketch book and I was constantly sketching, sketching ... at the very beginning I was quite scared and drawing very neatly – 'oh no I don't want to make a mistake' – but later on got more free and didn't really care. That is when I got my best bits – there was a point where I was really angry – I just couldn't get a design – and scribbled and 'oh actually that's quite good!' (Second-year industrial design student)

Introduction

A recent longitudinal study carried out by the Centre of Excellence for Product and Automotive Design (CEPAD) at Coventry University aimed to identify the crucial transformations that undergraduate industrial design students must inculcate in order to successfully join their global community of practice. The findings showed that the concept of a liminal space is an important factor.

The research, using the threshold concept theory developed by Meyer and Land (2003) followed a cohort of undergraduates from entry to graduation between 2005 and 2010: in total 89 students were interviewed during the lifetime of the research.

The threshold concept theory posits the idea that within disciplines there are conceptual gateways or portals, which – due to their troublesome nature – can make it difficult for students to progress. As such, a threshold concept is seen as distinct from 'core concepts' – or building blocks – within disciplines, due to the notion of transformation (Meyer and Land, 2006: 6). In other words, grasping a threshold concept will irrevocably transform a student's understanding, and this transformation can relate to the particular subject at hand, and/or be extrapolated beyond the academy.

A threshold concept is seen as a conceptual gateway, and is defined as:

> *akin to a portal, opening up a new and previously inaccessible way of thinking about something. [It] represents a transformed way of understanding, or interpreting, or viewing something without which the learner cannot progress. As a consequence of comprehending a threshold concept there may thus be a transformed internal view of subject matter, subject landscape, or even world view. (Meyer and Land, 2003: 1)*

Meyer and Land identified several possible characteristics of a threshold concept, with the first being 'transformative'. Consequently, understanding a threshold concept will result in a personal as well as conceptual change. As such this transformation will become part of whom the student is, how they see and feel (Cousin, 2006) and will therefore expand personal biographies.

Another characteristic is that a threshold concept is often 'irreversible', as once understood the learner is unlikely and often unable, to forget it.

A threshold concept can also be 'integrative', in that it opens up connections between different learning experiences and enables students to make conceptual leaps within a much wider playing field of knowledge: 'the landscape is different' (Meyer et al., 2008: 70).

Perhaps the most important characteristic of all is that of 'troublesome knowledge': 'knowledge that is conceptually difficult, counter-intuitive or "alien"' (Perkins, 1999 in Meyer and Land, 2003: 1). This is the characteristic that receives the most attention, and in essence, this is where students are required to move outside their comfort zone and enter sometimes disconcerting new territories. Meyer and Land go on to discuss how previous forms of knowledge need to be challenged in order to master a threshold concept.

Examples given are *ritual* knowledge – that which is routinely offered in response to a question, but which does not evidence the possible complex underpinnings of such knowledge; *inert* knowledge – which can be seen as 'stand-alone' and displays no interconnectedness with a wider context;

conceptually difficult knowledge – that which, if not grasped, leaves students unable to move from their intuitive knowledge, can result in mimicry of a subject and so troubled or limited understanding can occur; *alien* knowledge – that which is counter-intuitive to what students already think they know, and *tacit* knowledge – that which operates unseen and is often the background knowledge that informs particular disciplines or subject areas.

Threshold Concepts in Industrial Design

Since the theory was first posited in 2003, threshold concepts have been identified in disciplines as diverse as depreciation accounting (Lucas and Mladenovic, 2006); caring in health (Clouder, 2005); the concept of the 'other' in communication studies (Cousin, 2006); climate change in geography (Hall, 2011); hypothesis in biology (Taylor et al., 2012); opportunity cost in economics (Shanahan and Meyer, 2006); surface area-to-volume ratio in nano-science (Park and Light, 2010); personhood in philosophy (Cowart, 2010); disjunction and problem-based learning (Savin-Baden, 2006); Hallidayan rank scale in languages (Orsini-Jones, 2009) and central limit theorem in statistics/entropy in physics (Meyer and Land, 2005).

However, the work undertaken with the industrial design undergraduate students at Coventry University is unique, with the journey and results of the research reported in detail elsewhere (Osmond, 2012, Osmond and Turner, 2010, Osmond, Bull and Tovey, 2010, Osmond and Turner, 2008, Osmond, Turner and Land 2008).

In summary, a threshold concept was identified for first year undergraduates, namely *the toleration of design uncertainty*, defined as 'the moment when a student recognises that the uncertainty present when approaching a design brief is an essential, but at the same time routine, part of the design process'.

The toleration of design uncertainty can be mapped against the characteristics of threshold concepts developed by Meyer and Land as illustrated in Table 8.1:

Table 8.1   The toleration of design uncertainty. Updated from Osmond (2009)

Transformative	Students accept that the toleration of design uncertainty is the jumping off point to innovative design
Irreversible	This transformation incurs a cognitive shift in terms of students' design confidence
Integrative	Students recognise that everything they learn and experience is a legitimate source of inspiration – for example, accepting that those moments when they surface around thinking about subjects that are not directly related to the task may turn out to be the most important part of the process
Troublesome	Students accept that they will constantly experience and re-experience this 'surfacing around' as they hunt for a solution, even when they attain the status of professional designer

Liminal Spaces

Underpinning the characteristics of a threshold concept is the notion of *liminality*. Meyer and Land argue that while students are trying to grasp a threshold concept, they can remain 'stuck' as they oscillate between previous and new understandings, thus experiencing a disjunction, particularly in relation to problem-based learning (Savin Baden, 2000). Examples of such liminal or conceptual spaces include the period between adolescence and adulthood and first-time motherhood: once entered there 'can be no ultimate full return to the pre-liminal state' (Meyer and Land, 2005: 376).

This notion of liminality is reflected within the creativity literature. First, perhaps the most well-known definition of creativity is that of 'Eureka' moments, epitomised by the example of Archimedes in his bathtub. Perkins (2000) addresses this, but argues that the breakthrough in this case was not the result of a sudden realisation but the result of a process involving a 'long search', 'little apparent progress', 'precipitating event', and a 'cognitive snap and transformation' (p. 9). For our purposes, the liminal space in this instance is found in the long search and little apparent progress – here Perkins gives the examples of Archimedes struggling with Hiero's problem and da Vinci fussing 'endlessly with flight'.

Kleiman, with his 'creativity-as-process' theory, also touches on this when he argues that 'playing for the sake of playing' (2008: 213) is important to the creative process, as does De Bono (1995) when he advocates that both time and space are needed for creativity to flourish – in this case time and space to try on different-coloured hats, such as 'ideas and proposals' and 'evaluating the alternatives'. For Claxton – describing a creative approach called 'Thinking at the Edge' (TATE) – focus is important: 'learning the knack of delicate inward attention to a somatic process of epistemic evolution, in which hazy, pre-conceptual ideas are given time to unfold into novel forms of talking and thinking' (2006: 351).

Courage is also needed as argued by White when she cites Nickerson – 'timidity is not conducive to creativity' (2006: 436) and so the confidence to take risks is important. Further, she argues that teachers need to nurture their own creativity in order to enhance this within their students.

The time, space and courage needed in order to enhance creativity should be underpinned by a space of safety and structure, as discussed by Davies et al. (2013) in a review of educational research, policy and professional literature on teaching creativity in schools. The review results fell into three main categories – the physical environment, the pedagogical environment and external partnerships. As well as time to work without pressure, and opportunities for immersion in activities, the authors identified that within the pedagogical environment a combination of freedom and structure is needed in order to foster creativity:

> the provision of 'safe' structure appears to be particularly important to enable pupils to take risks, to think creatively and critically, and to question ... best served by an equal balance between structured and unstructured work. (2013: 85)

Second, and perhaps more importantly, the notion of a liminal space identified by the CEPAD research is also echoed in the design literature relating to professional designers. Thus the notion builds upon the work of design theorists such as Tovey (incubation period, 1984), Cross (oscillation between problem and solution, 1992), Dorst (tightrope walking, strategic thinking and visionary designer category, 2003); Wallace (problem bubbles, 1992) and Daly et al. (personal synthesis, directed creative exploration and freedom, 2012).

In more detail, Tovey's incubation period relates to right/left brain thinking, and he argues that it is only when both sides of the brain are in agreement after a period of incubation – where there is 'considerable interaction and interference' between each side – that the design process is 'concluded' (1984: 226).

For Cross, a similar process takes place, which he calls 'oscillation': 'designing seems to proceed by oscillating between sub-solution and sub-problem areas, as well as by decomposing the problem and combining sub-solutions' (2007: 78). In other words, Cross argues that preliminary models of the problem and solution can be found side by side in the brain where oscillation then takes place. Eventually when the two are reconciled – typically via a 'bridge across the chasm between problem and solution' – a creative leap takes place (2007: 78).

Dorst describes this process as 'tightrope walking', and notes that there is no certainty that a successful design will result, or how long this could take (2003: 97).

For Wallace, this process is put into the context of 'problem bubbles' within which designers jump from idea to idea (bubble to bubble). Stating that personal design thinking is not a 'linear flow', he argues that effective designers tend to immerse themselves within the bubbles, but at the same time have the ability to consciously 'hover above' them which enables both the determination of patterns and control over the process (1992: 80).

More recently, Daly et al.'s paper outlined findings from interviews with twenty professional designers from a range of design disciplines. The authors found that designers typically approach designing from six distinct design 'lenses' or categories. They argue that individual designers will emphasise one of the lenses within their repertoire: also that the approaches can be structured hierarchically (see Figure 8.1). Thus the 'results of this work reveal outcomes of how designers combine the skills, knowledge, and experiences they have with design and what aspects of design they emphasise' (Daly et al., 2012: 210).

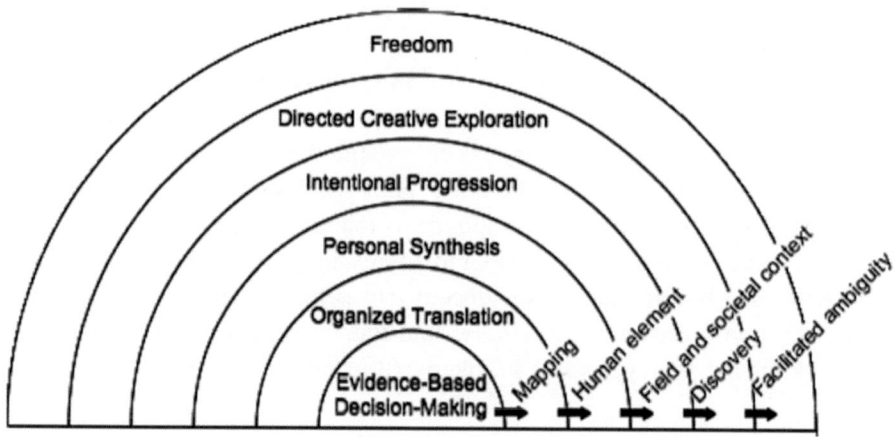

Figure 8.1   Outcome space in hierarchical form. Reproduced from *What Does it Mean to Design?* Shanna R. Daly, Robin S. Adams, George M. Bodner, 2012, John Wiley and Sons.

In more detail, the lenses/categories are described as follows (Table 8.2):

Table 8.2   **Six design lenses. Adapted from Daly et al., 2012: 198–204**

Category	Type	Description
Level 1 lens	Evidence-based decision-making	Designing from a basis of grounded evidence
Level 2 lens	Organised translation	Adds a consideration of the solution in terms of the end goal
Level 3 lens	Personal synthesis	Sees the designer as the conduit, bringing knowledge and experience to bear towards the finished design
Level 4 lens	Intentional progression	Includes acknowledgement of the temporal and future implications for the wider field within which the design will sit
Level 5 lens	Directed creative exploration	Recognises the need for flexibility, experimentation and possible changes in course
Level 6 lens	Freedom	Allows for facilitated ambiguity and limitless possibilities from beginning to end of the design task

As can be seen from the literature, the concept of a liminal space is well represented, and it is argued here that the toleration of design uncertainty as a threshold concept has resonance to three of Daly et al.'s design lenses, specifically, lens 3 (personal synthesis), 5 (directed creative exploration) and 6 (freedom). All of these levels speak to needing confidence – confidence to negotiate creativity and trust in one's personal judgement, confidence in the ability to be flexible, and confidence to tolerate ambiguity. Without this confidence each lens could be considered an unsafe space for student designers, who will often be meeting these concept lenses for the first time in creative courses that typically emphasise individual agency and are often underpinned by tacit knowledge and 'wicked problems' (Buchanan, 1992).

Thus, undergraduate students are encouraged to bring their creativity to the fore, perhaps through play and experiment.However, as novice designers, they will not possess the requisite skills, knowledge and experience – and most importantly the confidence – to successfully negotiate the attendant uncertainties inherent within these concept lenses, in particular in terms of the freedom aspect of lens 6.

Impact on the Curriculum

The need to provide safe spaces within the curriculum for students to experience intense uncertainty is recognised by Meyer and Land when they talk about the 'jewels in the curriculum' that encourage transformative moments in the student learning journey. Using these jewels as central points, the curriculum can offer a framework of engagement to promote conceptual understandings, and also can be used as diagnostic points for tutors.

Claiming that threshold concepts 'literally are the waypoints to be navigated … they are what really matters in the course and where the key transformations educators wish to bring about take place.' (Land and Meyer 2010: 75), the authors discuss how the liminal states that students enter as they approach the jewels in the curriculum are key points at which assessment practices can be used to assess how, and if, students have reached a point where they are able to inculcate a threshold concept.

This, they argue, is of utmost importance if assessment practices are to identify and address students who can 'produce the right answer, while retaining fundamental misconceptions' (2010: 62) and thus allow a moving away from the notion that students arrive at the 'finish line' of a course at the same time.

A 'one size fits all' assessment method will not capture the variations in student understanding (2010: 63–66). Variation in this case relates to the 'extent or degree to which individuals vary and perform understanding' (p. 64). The key to this is a recognition of liminal variation while students are suspended in the liminal space which, once recognised, can result in new and creative methods of assessment. This process will necessarily involve deviation and unexpected outcomes and so a course design that focuses on a prescribed outcome will preclude such a journey. Therefore, the threshold concept theory 'to some extent "rattles the cage" of a linear approach to curriculum design that assumes standard and homogenised outcomes' (Land et al., 2005: 59).

Consequently, the authors seek assessment practices that, to fully utilise the threshold concept theory, would offer ways of assessing student conceptual formations in four stages of liminality, possibly through an environment of 'rich' feedback at identified 'stuck' points (Land and Meyer 2010: 76). This should be underpinned with a clarification of variation within each liminal state and also a grading system that can cope with identified troublesome knowledge. Such a strategy would ideally need to take into account variation in student knowledge on entry to the course, how they approach the threshold concept, what coping strategies they employ when within a liminal space and how they move forward when emerging from the space.

In conclusion, Meyer and Land posit that the threshold concept framework enables a focus on the 'learning episodes' that facilitate understanding of transformative concepts (Land et al., 2005:70) .

The Industrial Design Curriculum

Therefore effective curriculum design is needed to provide safe spaces that allow students to iteratively struggle, fail and succeed if they are to successfully progress in their studies.

In 2010, the toleration of design uncertainty was used to redesign the Coventry University undergraduate industrial design curriculum for year one and year two to include such a safe space (see Tovey et al., 2010).

Separate quadruple practice modules for year one and year two which spanned each year were introduced with contained assessment gateways which assumed greater importance (in terms of marks) as the year progressed. Specifically the first assessment attracted 10 per cent of the total mark for the year, the second 15 per cent and the final 75 per cent. Accompanying this new curriculum model was the provision of dedicated spaces for the year one and year two students which they could colonise as their own.

The intention was to provide a safe, structured, creative space for industrial design students to allow them the time and space to immerse themselves within a series of design briefs, scaffolded by extensive formative feedback. The outcome, epitomised by the quote at the beginning of this chapter, indicated that the students were managing their anxiety within a supportive environment, and, as expected, there were some indications that students from a 'tick-box' educational background experienced some difficulty with accepting that it was OK to struggle, and sometimes fail, their initial assignments.

More recently, a teaching experience with a cohort of international MA students resulted in a team of tutors redesigning teaching delivery when it became apparent that a liminal space linked to the toleration of design uncertainty made its presence felt within a research methods module. For these students, the struggle occurred when they encountered ill-defined design problems during an initial teaching period which required them to expand their creative thinking from a solution to a problem focus. As reported in Bull and Osmond (2013) although the students understood the mechanics of creativity, they were unable to translate this in terms of a problem-solving context. This left the students in an uncomfortable liminal space as they concentrated on struggling to understand what the tutors wanted, rather than realising that they could experiment and play with ideas. As a result of identifying this, the tutor team recognised that the existing step-by-step delivery method had not helped the students to break out of their boundaried thinking processes and so they focused on a more conceptual delivery. The result was that the students slowly began to experiment with ideas, thus testing and expanding their own creative boundaries and building their design confidence.

Since 2010, other innovations in terms of curriculum design have been implemented in order to enhance students' creativity using the notion of threshold concepts and the concurrent liminal space. As reported in Johnson et al. (2013) an important threshold concept is that of cooperative learning. Often students enter the course with the mindset of the 'lone guy with the sketchpad' and from year one, the curriculum is now designed to address this in order to foster cooperative learning. Using some of the characteristics of threshold concepts, year one is considered to be *troublesome* in that students tend to think in an individual linear fashion. Thus, accepting that they need to work as a team and that their idea is not always the best one to go forward can be difficult. *Transformation* is the key phrase within year two, when there is further focus on cooperative work – internally by working on group projects and externally by working with external partners. By year three, the change in student thinking process is further encouraged through a mandatory work placement module. Students return from external work placements with the realisation that cooperative working is not optional and as such this realisation can be seen as *irreversible*. Finally, in year four, the *integration* of the previous years coalesces into routinely working as teams and becomes part of the student's identity as they prepare to enter their professional community of practice.

Conclusion

This chapter has outlined the important of the concept of liminal spaces in relation to an industrial design curriculum within higher education.

Based on the work of the Centre of Excellence for Product and Automotive Design (CEPAD), which identified the toleration of design uncertainty as a threshold concept, there is a consideration of the notion of liminal spaces in relation to both the creativity and design literature.

It is argued that for student designers, liminal spaces can be unsafe places because they will not have the skills, experiences and confidence necessary to negotiate them successfully, so a curriculum which first identifies its 'jewels' and second builds safe spaces around them can only enhance students' creative abilities.

Examples of curriculum changes within the industrial design course at Coventry University are given, most notably a major curriculum redesign for year one and two students, a module redesign for international MA students and a brief outline of how the threshold concept of cooperative learning is focused upon throughout the four year course.

In conclusion, there seems little doubt that giving students the time, space and structure to immerse themselves into a design brief can only enhance their creativity.

References

Buchanan, R. (1992) Wicked problems in design thinking, *Design Issues*, 8(2), 5–21.

Bull, K. and Osmond, J. (2013) Design education and non-EU students: shifts in teaching practice, in *DRS//CUMULUS Oslo 2013 Proceedings, Vol. 1*, pp. 357–370. Oslo, 14–17 May. Eds Reiten, J.B. et al. ABM Media, Oslo, Norway.

Claxton, G. (2006) Thinking at the edge: developing soft creativity, *Cambridge Journal of Education*, 36, 351–362.

Clouder, L. (2005) Caring as a threshold concept. Transforming students in higher education into health (care) professionals, *Teaching in Higher Education*, 10, 505–517.

Cousin, G. (2006) Threshold concepts, troublesome knowledge and emotional capital: an exploration into learning about others, in J.H.F. Meyer and R. Land (eds), *Overcoming Barriers to Student Understanding. Threshold Concepts and Troublesome Knowledge*, pp. 131–147. London and New York: Routledge.

Cowart, M. (2010) A preliminary framework for isolating and teaching threshold concepts in philosophy, in R. Land, J. Meyer and J. Smith (eds), *Threshold Concepts within the Disciplines*, pp. 131–145. Rotterdam: Sense Publishers.

Cross, N. (1992) Research in design thinking, in N. Cross, K. Dorst and N. Roozenburg (eds), *Research in Design Thinking*, pp. 3–10. Delft: Delft University Press.

Cross, N. (2006) *Designerly Ways of Knowing*. London: Springer-Verlag.

Daly, S., Adams, R. and Bodner, G. (2012) What does it mean to design? A qualitative investigation of design professionals' experience, *Journal of Engineering Education*, 101(2), 187–219.

Davies, D., Jindal-Snapeb, D., Colliera, C., Digbya, R., Haya, P. and Howea, H. (2013) Creative learning environments in education – a systematic literature review, *Thinking Skills and Creativity*, 8, 80–91.

De Bono, E. (1995) Exploring patterns of thought: serious creativity, *Journal for Quality and Participation*, 18(5), 12–18.

Dorst, K. (2003) *Understanding Design 150 Reflections on Being a Designer.* Amsterdam: BIS.

Hall, B. (2011) Threshold concepts and troublesome knowledge: towards a 'pedagogy of climate change'? Available from The Gees Subject Centre http://gees.ac.uk/pubs/other/pocc/pocc.htm, accessed 26 June 2013.

Johnson, C., Bull, K. and Osmond, J. (2013) Co-operative design and communities of practice. *The Tenth International Conference on Cooperative Design, Visualization and Engineering*, 22–25 September, Mallorca, Spain.

Kleiman, P. (2008) Towards transformation: conceptions of creativity in higher education, *Innovations in Education and Teaching International*, 45(3), 209–217.

Land, R. and Meyer, J. (2010) Threshold concepts and troublesome knowledge (5): dynamics of assessment, in R. Land, J. Meyer and J. Smith (eds), *Threshold Concepts within the Disciplines*, pp. 61–79. Rotterdam: Sense Publishers.

Land, R., Cousin, G., Meyer, J. and Davies, P. (2005) Threshold concepts and troublesome knowledge (3): implications for course design and evaluation, in C. Rust (ed.), *Improving Student Learning – Diversity and Inclusivity. Proceedings of the 12th Improving Student Learning Conference, 2004, Birmingham*, pp. 53–64. Available from http://www.brookes.ac.uk/services/ocsld/isl/isl2004/abstracts/conceptual_papers/ISL04-pp53-64-Land-et-al.pdf, accessed 19 June 2013.

Lucas, U. and Mladenovic, R. (2006) Dissolving the boundary between research and teaching: exploring threshold concepts within introductory accounting, *SRHE: Beyond Boundaries, New Horizons for Research into Higher Education*, 12–14 December, Brighton, UK.

Meyer, J. and Land, R. (2003) Threshold concepts and troublesome knowledge: linkages to ways of thinking and practising within the disciplines. Occasional Paper 4. Available from: http://www.etl.tla.ed.ac.uk/docs/ETLreport4.pdf, accessed 26 September 2013.

Meyer, J. and Land, R. (2005) Threshold concepts and troublesome knowledge (2): epistemological considerations and a conceptual framework for teaching and learning, *Higher Education*, 49, 373–388.

Meyer, J. and Land, R. (eds) (2006) *Overcoming Barriers to Student Understanding. Threshold Concepts and Troublesome Knowledge.* London and New York: Routledge.

Meyer, J., Land, R. and Davies, P. (2008) Threshold concepts and troublesome knowledge (4): issues of variation and variability, in R. Land, J. Meyer and J. Smith (eds), *Threshold Concepts Within the Disciplines*, pp. 59–74. Rotterdam: Sense Publishers.

Orsini-Jones, M. (2009) *Researching Learning in Higher Education.* London and New York: Routledge.

Osmond, J. (2009) Stuck in the bubble: identifying threshold concepts in design, in *Dialogues in Art and Design: Promoting and Sharing Excellence*, pp. 131–135. *GLAD Conference Proceedings*, 21–22 October at York St John University. Pub, ADM HEA Subject Centre, University of Brighton.

Osmond, J. (2012) Passports to a community of practice, in M. Tovey (ed.), *Design for Transport*, pp. 335–352. Farnham: Gower.

Osmond, J., Bull, K. and Tovey, M. (2009) Threshold concepts and the transport and product design curriculum, *Art, Design and Communication in Higher Education*, 8(2), 169–175.

Osmond, J. and Turner, A. (2008) Measuring the creative baseline in transport design education, in C. Rust (ed.), *Improving Student Learning – For What?*, pp. 87–101. Oxford: OCSLD.

Osmond, J. and Turner, A. (2010) The threshold concept journey: from identification to application, in J. Meyer, R. Land and C. Baillie (eds), *Threshold Concepts and Transformational Learning*, pp. 347–363. Rotterdam: Sense Publishers.

Osmond, J., Turner, A. and Land, R. (2008) Threshold concepts and spatial awareness in automotive design, in J. Meyer, R. Land and J. Smith (eds), *Threshold Concepts within the Disciplines*, pp. 243–258. Rotterdam: Sense Publishers.

Park, E. and Light, G. (2010) Identifying a potential threshold concept in nanoscience and technology: engaging theory in the service of practice, in

J. Meyer, R. Land and C. Baillie (eds), *Threshold Concepts and Transformational Learning*, pp. 259–279. Rotterdam: Sense Publishers.

Perkins, D. (1999) Threshold concepts and troublesome knowledge: linkages to ways of thinking and practising within the disciplines, in C. Rust (ed.), *Improving Student Learning Theory and Practice – 10 Years On*, pp. 412–424. Oxford: OCSLD.

Perkins, D. (2000) *The Eureka Affect: The Art and Logic of Breakthrough Thinking*. New York and London: W.W. Norton & Co.

Savin-Baden, M. (2000) *Problem-based Learning in Higher Education: Untold Stories*. Buckingham: SHRE and Oxford University Press. Available from: http://ww.kingscourt.co.uk/openup/chapters/033520337X.pdf, accessed 20 June 2013.

Shanahan, M. and Meyer, J. (2006) The troublesome nature of a threshold concepts in economics, in J. Meyer and R. Land (eds), *Overcoming Barriers to Student Understanding*, pp. 100–114. London and New York: Routledge.

Taylor, C., Liu, D., Pye, M., Tzioumis, V. and Meyer, J. (2012) Using threshold concepts to design a first year biology curriculum, in M. Sharma and A. Yeung (eds), *Australian Conference on Science and Mathematics Education*, p. 33. 26–28 September, University of Sydney. Published by: UniServe Science, The University of Sydney, NSW 2006, Australia.

Tovey, M. (1984) Designing with both halves of the brain, *Design Studies*, 5(4), 219–228.

Tovey, M., Bull, K. and Osmond, J. (2010) Developing a pedagogic framework for product and automotive design, in *Design Research Society Design & Complexity Conference*, July 7–9, Université de Montréal.

Wallace, K. (1992) Some observations on design thinking, in N. Cross, K. Dorst and N. Roozenbrug (eds), *Research in Design Thinking*, pp. 75–86. Delft: Delft University Press.

White, J. (2006) Arias of learning: creativity and performativity in Australian teacher education, *Cambridge Journal of Education*, 36(September), 435–453.

Chapter 9

Developing Tools to Support Collaboration and Understanding during Industrial Design Practice

MARK EVANS, IAN CAMPBELL AND EUJIN PEI

Introduction

Professional and student industrial designers employ an extensive range of media and techniques at various times during creative practice. While general patterns of use are acknowledged, such as loose sketches at the beginning of product development and full prototypes at the end, the nuances of use for specific design representations have been elusive, making contextualisation for students problematic. This chapter reports on research to enhance communication during product development by making tacit knowledge on the use of design representations explicit for both students and practitioners. This was achieved through the development of two design tools called CoLab and iD Cards. Phase 1 of the project identified barriers to communication through semi-structured interviews with 61 industrial designers and engineering designers at 17 industrial design consultancies. Phase 2 explored the nature of design representations and categorised 35 types as sketches, drawings, models or prototypes using semi-structured interviews with both industrial designers and engineering designers, with differences in use between the two groups becoming apparent. Phase 3 used a process of information design to translate the findings and data from Phase 2 into the card-based CoLab design tool that included the taxonomy and an indication of when the design representations were used by industrial designers and engineering designers and for what types of information. Changes were made after appraisal and the final tool was validated through semi-structured interviews with 43 industrial design and engineering design practitioners and observation. Phase 4 disseminated the research output with the support of the Royal Academy of Engineering (RAE) in the UK (CoLab web-based design tool) and Industrial Designers Society of America (IDSA) in the USA (iD Cards physical design tool). The chapter concludes that the use of appropriate research methods that integrate literature based sources with practitioner engagement has the potential to elicit valuable and unexpected tacit knowledge that can contribute to student learning. It also acknowledges that while the outcomes from such research can be enthusiastically received, translation into a format for effective dissemination can be a challenging and time-consuming process. However, with confidence in outcomes and a desire to disseminate, opportunities can be identified if researchers are prepared to be flexible and adapt to stakeholder needs.

Background

The complex and competitive nature of product development requires collaboration between design professionals to effectively conceptualise, develop and commercialise innovative products (Edmondson and Nemhard, 2009). Despite the importance of inter-disciplinary collaboration, few studies have examined the relationship between industrial design and engineering design. In the context of this study, industrial design is defined as the specification of product form and includes aesthetic judgement, semantics, user interface and social requirements (IDSA, 2006; Tovey, 1994; Flurscheim, 1983). In contrast, the term engineering design broadly encompasses mechanical, electrical and electronic engineering (Fielden, 1963), all of which employ science-based problem-solving methods (Hurst, 1999).

The aim of the research was to investigate problems associated with collaborative interaction between the communities of professional practice of industrial designers and engineering designers. Disharmony during NPD (New Product Development) may occur when team members approach a project differently. For example, industrial designers adopt open-ended solutions, using instinct and trial and error to embody personal creativity for the design; while engineering designers view problems as precise and focus on functionality, specification and performance (Kim and Philpott, 2006). In terms of deliverables, engineering designers produce technical details for manufacture, based on quality, performance and cost (Flurscheim, 1983); while industrial designers deliver visual representations such as sketches and physical models. As a result, their dissimilar views and contrasting outcomes may create conflict (Persson, 2002).

Previous research has focused on inter-disciplinary collaboration between engineering design and manufacturing (Beskow, 1997; Ulrich and Eppinger, 2000) and engineering with marketing (Griffin and Hauser, 1996; Shaw and Shaw, 1998). With the exception of Persson and Warell (2003), who identified methods and tools adopted by industrial designers and engineering designers, research to investigate the collaborative interaction between industrial designers and engineering designers is under-represented. Persson and Warell (2003) reported that communication, social factors, personality differences and physical settings were key factors in influencing professional interaction. Persson (2005) went on to propose a collaborative workspace with a joint mindset by means of socialisation and mediating instruments to enhance collaboration. Other integrating mechanisms included social organisation (Kahn, 1996; Jassawalla and Sashittal, 1998), the use of inter-communal negotiation for better cross-functional teamwork (Brown and Duguid, 2001), having boundary-spanning and good teaming skills (Edmondson and Nemhard, 2009), and employing information and communication technology (Sproull and Kiesler, 1991; Toye et al., 1993). Although other established methods, such as Quality Function Deployment (QFD) and stage-gate solutions are available (Ulrich and Eppinger, 2000), they are primarily designed for engineers. As such, very few integrating mechanisms are available to enable, facilitate or improve collaboration been industrial designers and engineering designers.

Rothwell (1992) proposed that effective communication and cross-functional linkages are the primary factors for successful NPD. Communication can be made effective by transmitting symbols precisely, ensuring that the meaning is relayed correctly, receiving the intended meaning accurately, and reaching the right audience through proper distribution (Chiu, 2002). Although communication mechanisms exist, researchers have observed that industrial designers and engineering designers still do not fully understand each other (Fiske, 1998). Communication only becomes accurate and effective when the team develops a common vocabulary and by understanding the communicative codes and language within the message content (Persson and Warell, 2003). In addition, collaboration represents a higher-level relationship when compared to communication that is limited to information exchange. Jassawalla and Sashittal (1998) stated that collaboration occurs when participants command equal interest, adopt transparency with high awareness, are mindful through integrated understanding, and perform with synergy. Collaboration allows members from different teams to divide work effectively, assist each other in maximising their joint contribution, and communicating accurate information such as through the use of precise design representations.

In the context of an opportunity to enhance collaboration between industrial designers and engineering designers by standardising language, developing awareness of methods and identifying differences in the use of design representations, the authors defined a methodology that would generate a taxonomy of design representations and then be used to collect empirical data that would confirm accuracy and identify when they were used and for what types of information by the two groups. By standardising language and providing a level of understanding of how industrial designers and engineering designers use design representations, a knowledge framework would be generated with the potential to translate into some form of design tool.

Phase 1: Identification of Barriers to Communication

INTERVIEWS

Semi-structured interviews were undertaken with experienced industrial designers, engineering designers and design managers from 17 industrial design consultancies specialising in consumer electronic products. There was a balance of large (more than 10 design staff), medium (between 6–10 design staff) and small industrial design consultancies (less than 5 designers) to allow a wider sampling and to obtain findings from a larger pool of respondents. Sixty-one semi-structured interviews were conducted. A semi-structured interview was selected because this method had the capacity to explore issues with the potential for respondents to fully describe personal experiences relating to group interaction and inter-disciplinary collaboration. After gathering general demographic data (educational background, work experience and the company structure) the participants were asked project-specific questions to identify factors relating to collaborative work. This required an example of a project, experiences of group interaction, reasons for project successes and failures, and an indication of the tools and methods used for the project. The questions can be seen in Table 9.1.

The interviews identified issues relating to inter-disciplinary collaboration which were encoded into a spreadsheet. A coding and clustering technique was then used to analyse the qualitative data and to help build theory (Miles and Huberman, 1994), as well as reducing data into themes and relationships (Strauss and Corbin, 1990). This pattern coding has been used by other researchers (Purcell and Gero, 1996) in order to summarise findings into condensed categories. The issues were reorganised with the most frequently occurring problems in a descending order as shown on the right column of the chart in Table 9.2.

Table 9.1   Questions used during semi-structured interviews

Research-specific questions
1. Describe a recent project undertaken
2. Describe the design approach and strategy adopted
3. What was the project deliverable?
4. What activities were involved?
5. Describe the tools and methods used
6. What design representation methods were used?
7. Did collaboration between industrial designers and engineering designers occur during the project?
8. Describe the quality of group interaction and teamwork
9. What factors might have influenced group work?
10. Were there any leadership or management issues?
11. Name the success or failure factors
12. What is your view of the final product?
13. Did you have any personal concerns working with the other discipline?
14. Suggest some improvements for future collaborative work

OBSERVATION

Following the interviews, observations were conducted to obtain detailed information by being close to the field of study. The use of observations is advantageous because it allows the researcher to examine interaction taking place between engineering designers and industrial designers in their natural working environment and to record potential barriers that might have occurred. The observations took place through a commercial design project over two consecutive weeks and involved the design of a consumer product with an industrial designer and an engineering designer working together. The observation was conducted at a design consultancy within its normal work environment and took place from the beginning of the project (design briefing) to the embodiment stage (3D CAD modelling). As video and voice recordings were not allowed due to project confidentiality, note-taking was used because it allowed conversations to be recorded and enabled first-hand accounts of the interaction to be documented. Reliability was achieved by cross-checking records during breaks to minimise work disruption. Other documents, including reports, specification lists and physical or virtual artefacts provided a more complete understanding of the design activities. To obtain a holistic view of issues within the project, observations were undertaken with the project leader, industrial designer and engineering designer.

Table 9.2   **Matrix of 61 problem categories tabulated from interviews**

#	Issues	1	2	3	4	5	6	7	8	9	10	11	12	13	14	15	16	17	Occurances	Category
1	Having knowledge of the other field				■	■	■		■		■				■	■		■	8	A
2	Conflict in Principles	■	■			■		■		■					■				6	A
3	Choosing the right tools and methods	■	■					■			■				■			■	6	B
4	Communication Skills	■				■			■						■	■		■	6	B
5	Use of Representation						■		■		■				■			■	6	B
6	Understanding each other					■		■		■					■			■	5	A
7	Fixed Engineering Mindset	■						■		■					■			■	5	C
8	Individual Differences & Attitude							■			■			■	■			■	5	C
9	Direction of Project Manager								■		■		■		■			■	5	A
10	Use of Rapid Prototype for Representation					■			■		■				■				4	B
11	Designers and Engineers having Different Values					■		■		■					■				4	C
12	Having a Common Goal								■						■	■			3	A
13	Get-together updates / Milestones									■					■	■			3	B
14	Informal Meetings	■							■						■				3	A
15	Understanding through Experience							■			■				■				3	C
16	Translation from 2D to 3D								■						■	■			3	B
17	Company Emphasis on Design or Engineering						■		■						■				3	A
18	Educational Background of Individual				■						■				■				3	C
19	Western vs Asian approach of working	■									■				■				3	C
20	Conflict in Interest	■									■								2	-
21	Fixed Working Protocols									■					■				2	-
22	Location of support members										■				■				2	-
23	Trust as a high-level understanding							■					■						2	-
24	Knowing the technical requirements							■		■									2	-
25	Working towards Joint-Solutions							■			■								2	-
26	Production & Manufacturing Limitations							■		■									2	-
27	Company Culture	■									■								2	-
28	Engineers do not Understand Role of Designers													■	■				2	-
29	Teamworking & Team Dynamics						■								■				2	-
30	Having standard Computer files						■								■				2	-
31	Limitations in Time leading to Poor Engineering												■		■				2	-
32	Limitations to size of Electronic Components								■	■									2	-
33	Creativity and Flexibility of Engineer					■								■					2	-
34	Marketing controls Budget affecting Design Quality			■										■					2	-
35	Language as a Probable Barrier					■			■										2	-
36	Knowing who is in charge / Roles & Responsibilities											■		■					2	-
37	Team Dynamics														■				1	-
38	Being specific				■														1	-
39	Designers getting carried away & fall behind time												■						1	-
40	Using standard codes														■				1	-
41	Having Multi-cultural Teams															■			1	-
42	Having Multi-disciplinary Teams															■			1	-
43	Fostering Team-spirit													■					1	-
44	Complexity of Project														■				1	-
45	Marketing Understand Designers Working					■													1	-
46	Designers Understand Manufacturing Constrains										■								1	-
47	Testing, Reviewing, Changing, Refining													■					1	-
48	Marketing should be faster to React													■					1	-
49	Engineering Issues affecting Design Aesthetics										■								1	-
50	Client Changes affecting Design Process					■													1	-
51	Designers not understanding Marketing Viewpoint	■																	1	-
52	Trimming Cost affecting Design Aesthetics							■											1	-
53	Difficulty in Explaining visual effects to Engineers													■					1	-
54	How Company & Organization Values each field															■			1	-
55	Software Incompetence															■			1	-
56	Proper justification for each decision to Understand																■		1	-
57	Using Technology for Enhanced Communication											■							1	-
58	Changes in Design due to Safety Requirements						■												1	-
59	Client Involvement in Design Stage							■											1	-
60	Education as a means to close gap btw Eng & Des								■										1	-
61	Difference between a Designer and Artist			■															1	-

The observations identified that formal and informal meetings were extremely valuable in enhancing collaboration. Co-location was an important factor since both industrial designers and engineering designers were located close to each other and had significant interaction when compared to other departments who were on a different floor in the building. The observations recorded different working approaches in which engineering designers focused on technical properties and cost whereas industrial designers explored form and expression. In addition, the lack of a common language in design representations caused miscommunication where certain words were interpreted incorrectly. For example, the engineers had intended simple sketches but the designers interpreted their task as requiring renderings which the engineers regarded as time-consuming and unnecessary at that stage. The generic term 'sketch' did not fully describe the requirements and deliverables for both parties. The observations also found that the loosely rendered sketches from the industrial designers were imprecise and the elliptical shapes drawn in perspective became hard to translate into a three-dimensional solid in CAD.

OUTCOMES FROM INTERVIEWS AND OBSERVATIONS

The interview study identified three problem areas in collaborative design which related to conflicts in values and principles. The first, conflicts in values and principles, related to the fact that engineering designers worked systematically based on quantified solutions. In contrast, industrial designers favoured an open-ended approach and used open solutions. The second, differences in design representations, noted that engineering designers often used technical terms and facts that included calculations, technical information and specifications; whereas industrial designers used freehand sketches and drawings to communicate ideas. The third, differences in education, was due to the fact that engineering designers were taught to employ systematic problem-solving and to justify solutions with facts. In contrast, industrial designers were taught to solve problems intuitively, rarely relying on quantified data. Due to differences in their educational background, both professions had different specialisations, approaches and expectations.

In addition, the observations revealed the significance of formal and informal meetings; the importance of co-located members; and the issue of having different interpretations of design representation terminology. Of these, the problem area of design representations was found to be highly significant in both interviews and observations and a decision was made to conduct a further investigation.

Phase 2: Investigating the Use of Design Representations

This phase explored the nature of design representations, generated a taxonomy, and collected data on when they were used and for what types of information.

THE NATURE OF DESIGN REPRESENTATIONS

The problematic nature of the use of design representations during product development necessitated an in-depth examination of their nature and function during product development.

Design representations can be expressed through language, graphic or artefacts (Goel, 1995; Goldschmidt, 1997) and they refer to models of the object being symbolised (Palmer, 1987). During the early stages of product development, representations such as sketches tend to be quickly produced and are relatively unstructured. As the design develops, more controlled methods such as drawings and models tend to be employed. Leonard-Barton (1991) noted that the progression of having more information embedded within a representation enhances the understanding of the design. For the practicing designer, sketches support visualisation, communication and information storage (Tang, 1991); externalising ideas (Larkin and Simon, 1987); thinking (Suwa et al., 1998); verification of decisions and allow a range of interpretations for a design solution (Scrivener, 2000).

While many forms of design representations are available, sketching is seen as being central during the early stages of product development. Goel (1995) sees sketches as the first step of the design process to externalise and visualise ideas at an individual level. At the next stage, representations are used to communicate with others and include presentation drawings and physical models. In the later stages, detailed technical drawings and prototypes are used to communicate detail. In comparing the differences between the representations favoured by industrial designers and engineering designers, Veveris (1994) observed that engineering designers used models associated with engineering principles, functional mechanisms, and production issues; whereas industrial designers applied representations related to appearance and usability. Despite various attempts to classify representations (Tjalve et al., 1979; Ullman et al, 1988; Tovey, 1989; Evans, 1992;; Veveris, 1994; Kavakli et al.,Tovey 1997 1998; Cross, 1999; Do et al., 2000; Otto and Wood, 2001; Cain, 2005; Olofsson and Sjölén 2005; Pavel 2005; Pipes 2007; Eissen and Steur 2008), they are largely incomplete or do not incorporate both industrial design and engineering design representations. In addition, researchers have noted problems with their use when symbolic elements become unclear. The more incomplete or vague a representation is, the greater and wider the perceptual interpretation space becomes. Despite such drawbacks, ambiguous representations allow for creativity and the generation of open-ended solutions (Rodriguez, 1992; Ehrlenspiel and Dylla, 1993; Fish, 1996). They enable things to be seen in different ways that in turn produces new designs and allows flexibility in terms of design attributes.

Although ambiguous representations possess benefits, their ill-defined nature makes it difficult for engineering designers to comprehend and recognise how they work in relation to a product's technical parameters (Saddler, 2001). It may be difficult for a viewer other than the originator to understand the embodied meaning, context or scale (McGown et al., 1998). The need for accurate and effective representations has been shown by Stacey and Eckert (2003) who provided an example of confusing sketches used in the knitwear industry. They cited that although the lines of a garment sketch were intended to describe the structure pattern, they could be misinterpreted as being stripes on the fabric.

TAXONOMY OF DESIGN REPRESENTATIONS

Following a comprehensive literature review to identify the key design representations used during product development and the information they were used to communicate, a taxonomy was generated that categorised 35 design representations as sketches, drawings, models and prototypes. Eighteen types of information that the design representations were used to communicate were also identified, being categorised under the headings of 'Design information' and 'Technical information'. The categories can be seen in Table 9.3.

Table 9.3   **Categories of sketch, drawing, model, prototype and categories of design information and technical information**

Sketches	Idea sketch	Information sketch
	Study sketch	Renderings
	Referential sketch	Inspiration sketch
	Memory sketch	Prescriptive sketch
	Coded sketch	–
Drawings	Concept drawings	Multi-view drawing
	Presentation drawing	General arrangement drawing
	Scenario and storyboard	Technical drawing
	Diagram	Technical illustration
	Single-view drawing	–
Models	3D sketch model	Concept of operation model
	Design development model	Production concept model
	Appearance model	Assembly concept model
	Functional concept model	Service concept model
Prototypes	Appearance prototype	System prototype
	Alpha prototype	Final hardware prototype
	Beta prototype	Tooling prototype
	Pre-production prototype	Off-tool prototype
	Experimental prototype	–
Design information	Design intent	Single views
	Form and detail	Multi views
	Visual character	Areas of concern
	Usability and operation	Texture and surface finish
	Scenario of use	Colour
Technical information	Dimensions	Mechanism
	Construction	Part and section profile lines
	Assembly	Exploded views
	Components	Material

DATA COLLECTION ON THE USE OF DESIGN REPRESENTATIONS

The taxonomy and categories of information were translated into matrices to use as research instruments (Pei et al, 2011). The first matrix was used to appraise the categories of the taxonomy and collect data on when the representations were used. The interview structure and process was identical to that of the first stage of interviews and involved 27 participants, of which there were 13 industrial designers, 10 engineering designers and 4 project managers. The results indicated that industrial designers employ sketches and engineering designers prototypes. While engineering designers did sketch, this tended to be during concept generations but industrial designers employed this during the entire process. The second matrix (see Figures 9.1, 9.2, 9.3 and 9.4) investigated the types of design and technical information that were present within sketches, drawings, models and prototypes.

The findings from the second matrix-based survey indicated that sketches, drawings and models provided a balanced range of design and technical information, with prototypes focusing on technical information. It was also apparent that design information was more commonly used by industrial designers than engineering designers. Conversely, technical information was more commonly used by engineering designers.

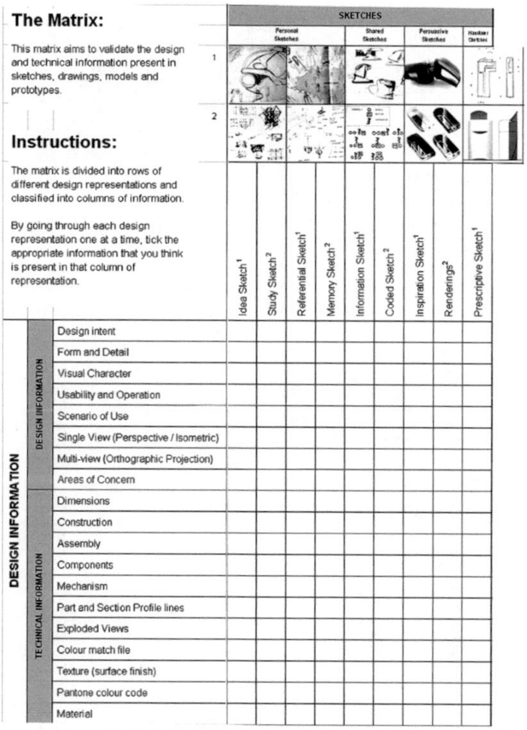

Figure 9.1   Design and technical information present in sketches

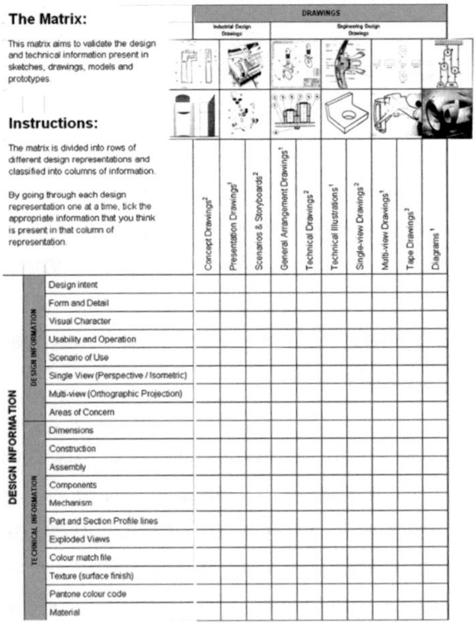

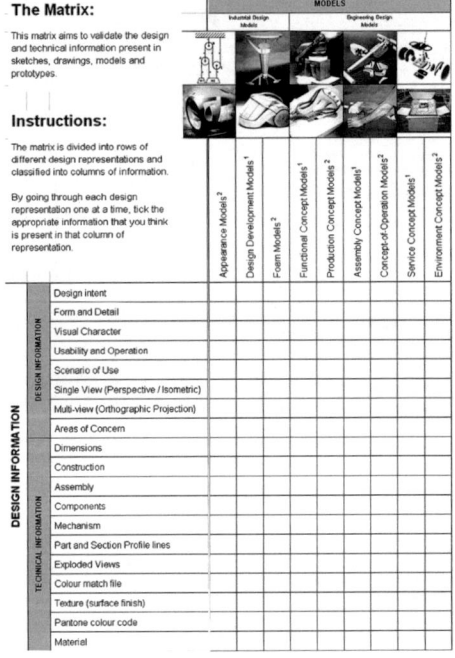

Figure 9.2   Design and technical information present in drawings

Figure 9.3   Design and technical information present in models

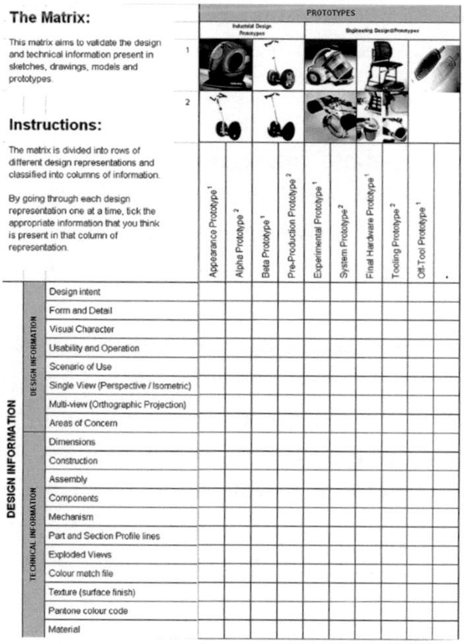

Figure 9.4   Design and technical information present in prototypes

Phase 3: Development of Design Tool

Having defined a taxonomy of design representations and collected data to identify the different way in which industrial designers and engineering designers, Phase 3 translated this knowledge framework in to a useable design tool (Pei et al, 2010). For the development of the design tool, several factors were used to determine the tool specification. According to Saddler (2001), the industrial design profession has representations that are ill-defined, imprecise and lack in communicative power. In addition, communication could be improved by having a common understanding of shared definitions (Matthew, 1997). Therefore, the primary feature of the design tool was to clarify the terminology of design representations and to act as an effective means of communicating these shared definitions. To meet this requirement, several physical formats were developed, including matrices, flowcharts, wheel diagrams and Rolodex systems. Digital formats were also considered but this meant that users would need to have constant access to a computer and it would be impractical to carry a laptop at all times. While personal digital assistants, tablets or smartphones presented more portable options, the dissimilar operating systems, short battery life and small screens would create additional problems for information retrieval. In addition, Wi-Fi or Internet-based tools would be limited to subscribers or connectivity.

Following an appraisal by the authors, a physical card format was selected for portability and immediate interaction between users. The aim was for the cards to be used by industrial designers and engineering designers as a portable tool that could be carried around as a reference guide or kept as an office resource or learning tool.

The design was undertaken by the researchers and, after numerous iterations, the knowledge framework was translated into 2 sets of 57 cards each. Both sets of cards included an identical taxonomy but differed in that there was a red set for industrial designers (with information on when this group used the design representations in the taxonomy and for what types of information) and a blue set with similar information dedicated for engineering designers. The principle behind the cards was to standardise language and demonstrate differences in the use of design representations by each group (Pei, 2009). Each pack comprised 4 cards describing the 4 design stages of product development (Set 1); 10 design information cards plus 8 technical information cards (Set 2); and 35 design representations (Set 3) of the taxonomy. Cards for an Idea sketch in the taxonomy section can be seen in Figure 9.5, with the bar graphs indicating what this was typically used for and when, with the industrial designer card being red and engineering designer blue.

APPRAISAL

The design tool was appraised through a pilot study that involved interviews with 10 design practitioners. Feedback indicated that a numerical referencing system would support faster access to information and a larger card format (ISO B8 size of 62 × 88 mm) would improve readability. Other improvements include a simplified layout with less text and larger images. These changes were implemented and the background redesigned to reduce the visual clutter. The revised design for is shown for the entire Sketches section of the taxonomy in Figure 9.6.

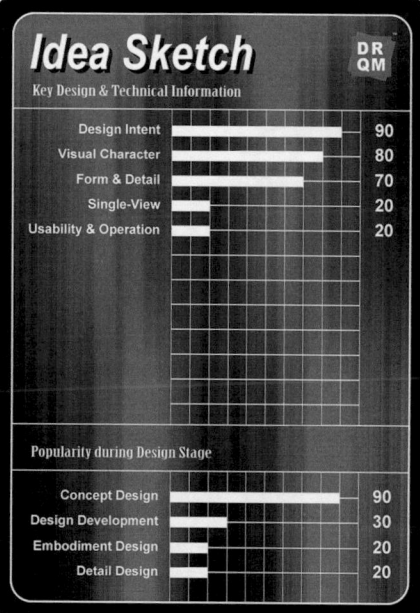
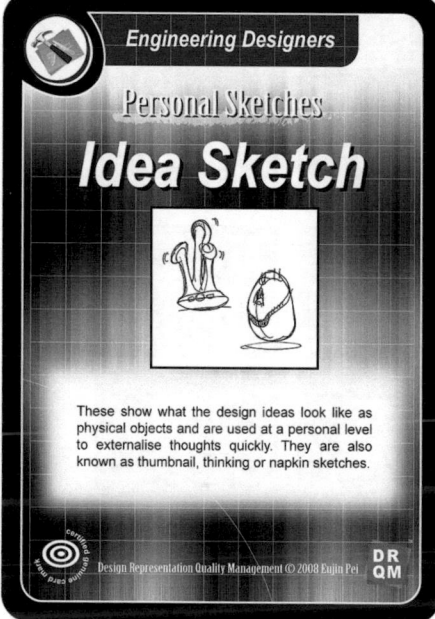
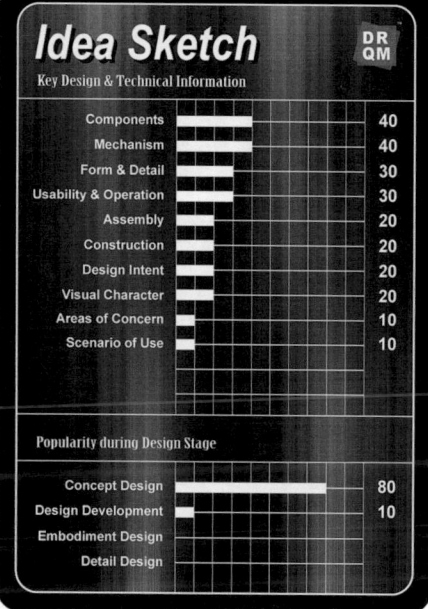

Figure 9.5   Idea sketch cards for design tool

Figure 9.6   Revised version of the cards for the sketches section of the taxonomy

VALIDATION

Having integrated several revisions, validation was undertaken through semi-structured interviews with final year industrial design (× 4) and engineering design (× 14) undergraduates who had worked together on an industrial project; experienced practitioners (× 43); and an observation study (× 1) to identify the contribution when the cards were used during the design of a consumer product in an industrial design consultancy.

In the student interviews, all industrial design students and 92.9 per cent of the engineering design students provided 'good' and 'excellent' feedback on the physical format of the cards. All industrial design students and 85.5 per cent of engineering design students felt that the tool would provide an enhanced understanding of design representations. 66.7 per cent of industrial design students and 64.3 per cent of the engineering design students felt that the cards would be effective in creating common understanding of design representations. While some students found it relatively difficult to search for the correct card, if a systematic approach was followed this should not have been a problem. A significant finding from the interviews was that all industrial design students and 85.8 per cent of engineering design students felt that the tool would have helped to foster enhanced collaboration.

All 43 of the practitioner participants were presented with identical questions to those of the students. When asked about the physical format, 86.4 per cent of industrial designers and 89.5 per cent of engineering designers gave a good/excellent rating. They also believed that that the tool would provide an enhanced understanding and clearer definition of design representations, with 86.4 per cent of the industrial designers and 89.5 per cent of the engineering designers offering agreement. In terms of the capacity of the cards to create a common understanding of design representations, 86.4 per cent of industrial designers and 84.2 per cent of engineering designers believed that they would achieve this. When asked if the system would foster enhanced collaboration, 68.2 per cent of industrial designers gave a good/excellent rating and 27.3 per cent were neutral: 63.2 per cent of the engineering designers gave a good/excellent rating and 36.8 per cent were neutral. A small number of participants claimed that experienced practitioners did not need these cards.

The observation study involved the design of a consumer product within a consultancy over a three-week period. Observing how the tool was used within a commercial context proved to be an extremely useful exercise because the authors could not predict how the tool would be received during practice. The industrial designers, engineering designers and team leader were observed and interviewed at the end of each day. During the observations, it was noted that the cards were useful as a clarification tool during the design process. On commencement of the third week, it became apparent that both industrial designers and engineering designers used identical keywords that had been learnt from the cards, thereby minimising the potential for misunderstanding. For example, the engineering designer started to request a more specific type of representation as opposed to a 'sketch' as a generic term which enabled more precise and relevant representations to be delivered. Similarly, when there was a need for a specific type of technical information, the industrial designer would refer to the cards to find the exact design representation that was required. The findings from the observations reinforced results from the interviews and provided further evidence of the potential for the tool to foster collaboration in a multidisciplinary environment.

The validation indicated that most participants gave an excellent and good rating for the design tool although it must be acknowledged that the sample size was limited to 65 participants.

Phase 4: Dissemination and Impact

The overwhelmingly positive response to the CoLab tool indicated the contribution and value of the cards. As academic research, it would have been possible for the researchers to have concluded their work at the validation stage but a decision was made to maximise impact through a process of dissemination.

COLAB

Despite making contact with numerous commercial and non-profit organisations who saw value in the CoLab tool, the relatively expensive production costs for 114 double-sided playing-card-size cards was prohibitively high. While the researchers had made a conscious and informed decision to create a physical design tool, there was a fundamental change in direction when an opportunity arose to translate CoLab into a web based tool with the support of the UK's Royal Academy of Engineering.

With funding from the National HE STEM Programme, the data on when design representations were used and for what types of information was translated into a database-driven website with functionality that was almost identical to that of the physical CoLab cards.

The orange tab at the bottom of the card indicates that it is one of the four Design Stages cards and the white background shows that Embodiment Design has been selected. The red card shows that the most popular design representation used by industrial designers during this stage was the Multi-view Drawing (70 per cent) and the blue card showing the most popular design representation used by engineering designers during this stage was the Technical Drawing (70 per cent). An additional level of functionality enables the user to click on the wording for the Design Representation and this then reveals full details on that particular card.

The purple tab at the bottom of the card indicates that this is from the Sketches section of the Design Representations. The red card indicates that Idea Sketches were mainly used by industrial designers (90 per cent) to provide information on Design Intent and the blue card by engineering designers (50 per cent) to provide information on Components.

The CoLab website is available on Loughborough University's Design Practice Research Group web site http://www.colab.lboro.ac.uk

ID CARDS

As a member of the Industrial Designers Society of America (IDSA), one of the researchers had presented the development of the tool at several of their international conferences. Following significant interest, particularly for the taxonomy, information on the use of design representations by industrial designers only was translated by two of the researchers into a fold-out tool using the Z Card printing process. This enabled the folds created by the 48 credit card-sized panels to replicate the card-based approach (see Figure 9.7).

The revised tool, called iD Cards, was approved by the Board of Directors of the IDSA in January 2011 and 5000 sets ordered for distribution to their practitioner, educator and student members. Further validation of the contribution of the iD Cards was received when they became a finalist in the 2011 International Design Excellence Awards. The information provided in the iD Cards group representations as Sketches, Drawings, Models and Prototypes, indicating when an individual card is used (yellow tab active) and for what type of information (red tab active for a type of Design Information, blue tab active for a type of Technical Information). Details in the type of information is provided on a separate panel as are instructions on use. The two sides of the folded-out iD Cards can be seen in Figure 9.8 and 9.9. and a close up image of the Idea Sketch is shown in Figure 9.10.

Despite the researchers wish to create a physical design tool, ongoing demand necessitated the conversion of the iD Cards into a pdf that was made available on Loughborough University's Design Practice Research Group website, http://www.lboro.ac.uk/media/wwwlboroacuk/content/lds/downloads/research/researchgroups/designpractice/IDSA%20iD%20Cards.pdf

Figure 9.7   iD cards
Source: Courtesy of Loughborough University and Eujin Pei

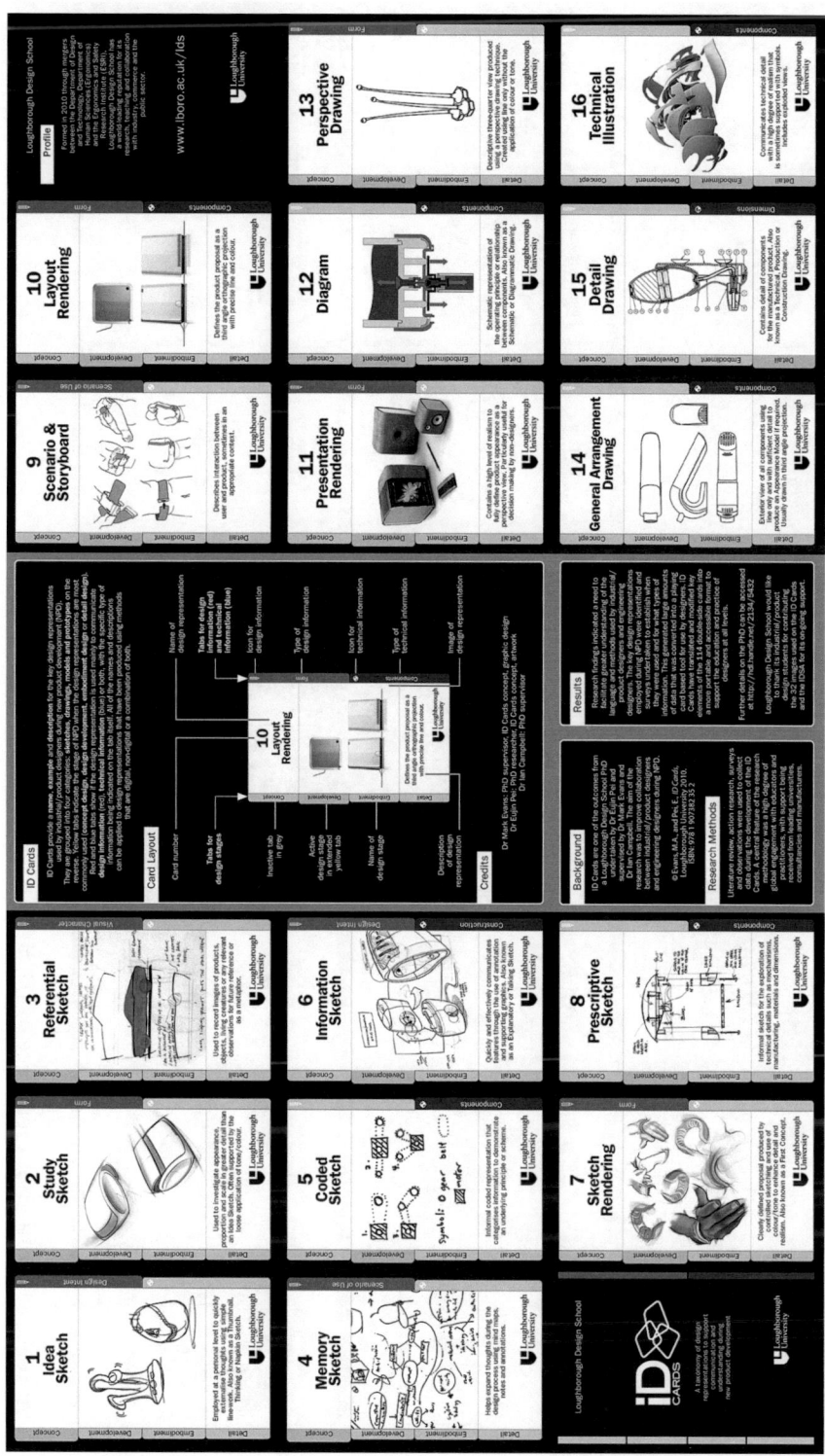

Figure 9.8   Folded-out front face of iD cards

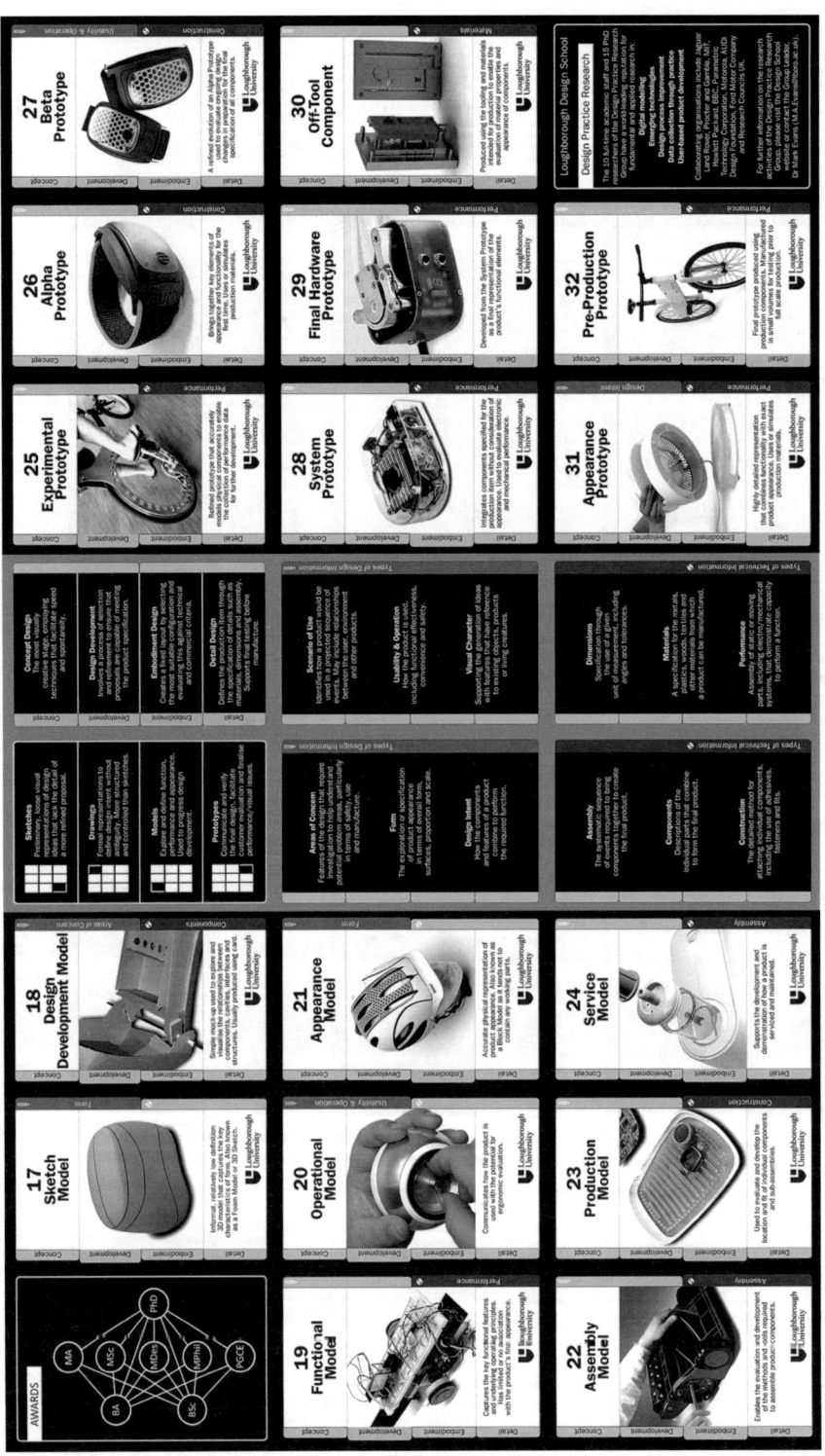

Figure 9.9   Folded-out rear face of iD cards

Figure 9.10   Idea sketch card on iD cards

Implications of CoLab and iD Cards for Design Education

While extensive teaching material is available to support students in the development of their craft/technical ability to create sketches, drawings, models and prototypes, the key published texts use a variety of descriptive terms which can have a negative impact on comprehension due to lack of standardisation. Both CoLab and iD Cards help to resolve this by providing names that are supported by descriptions (some of which include alternative names) and images. The impact of CoLab and iD Cards is then extended by contextualising the design representations during product development which enables students to identify timing and appropriateness. This is particularly relevant when students have a temptation to offset their under-developed sketching skills (typically used during the Concept Design stage) by using 3D CAD to visualise ideas which is more appropriate during the Design Development stage. CoLab and iD Cards help identify such discrepancies and enable reflection and the presentation of alternative approaches.

CoLab and iD Cards do not have a defined position in the industrial design curriculum as, by their nature, they offer generic information. By making information readily available that names the key design representations, when they are used and for what types of information; there is considerable scope for educators to integrate the resources into their teaching as required. For example, they could be used to support the development of sketching capability by providing curriculum guidance on the craft skills required to develop competence in the generation of each type of sketch (Idea Sketch, Study Sketch etc.). Alternatively, as CoLab and iD Cards indicate when the design representations are used during product development, their contribution extends to design management and strategies for effective client communication. Collaboration with engineering designers can then be explored further by using CoLab to identify the different ways design representations are used as compared to industrial designers.

The positive response of educators and students to CoLab and iD Cards indicates a key role in the use of academic research to provide a theoretical link with practitioner activity that is grounded in an empirical context.

Conclusions

Design representations are an integral component of product development because they support innovation through the externalisation, manipulation and communication of design. The fact that design representation, collaboration and communication are closely linked means that the use of CoLab and iD Cards can contribute to student learning and professional practice by presenting a language platform to standardise vocabulary, thereby facilitating social networks and enhancing understanding between stakeholders. The context where the tools can be used is not limited to student and practitioner industrial designers and engineering designers but has the potential for use by commercial stakeholders, including marketing and production engineering. Additionally, the tool has an application as a teaching and learning tool in design education.

While the formalisation embodied in the tool might be seen as introducing rules and procedures which, at times, may have a negative impact (Burns and Stalker, 1961), the authors believe that a focused system can minimise misinterpretation and lead to more accurate communication. By including key design and technical information, the tools serve as decision-making guides and help identify representations used during design stages. They also allow industrial designers and engineering designers to be aware of each others' working practice and aid the coordination of actions, task management and the anticipation of actions by others (Gutwin and Greenberg, 1996). Through the use of the CoLab and iD Cards, inter-disciplinary teams are able to develop a shared language to communicate effectively and students become familiar with such working practices.

References

Beskow, C. (1997) Product development in change – cross-functional co-operation and pdm implementations, Licentiate Thesis. Department of Machine Design, Royal Institute of Technology, Stockholm.

Brown, J.S. and Duguid, P. (2001) Knowledge and organization: a social-practice perspective, *Organisation Science*, 12(2), 198–213.

Burns, T. and Stalker, G.M. (1961) *The Management of Innovation*. London: Tavistock.

Cain, R. (2005) Involving users in the design process: the role of product representations in co-designing, Ph.D. Thesis. Department of Design and Technology, Loughborough University: Loughborough.

Chiu, M.-L. (2002) An organizational view of design communication in design collaboration, *Design Studies*, 23(2), 187–210.

Cross, N. (1999) Natural intelligence in design, *Design Studies*, 20(1), 25–39.

Do, E., Yi, L., Gross, M.D., Neiman, B. and Zimring, C. (2000) Intentions in and relations among design drawings, *Design Studies*, 21(5), 483–503.

Edmondson, A.C. and Nembhard, I.M. (2009) Product development and learning in project teams: the challenges are the benefits, *Journal of Product Innovation Management*, 26(2), 123–138.

Ehrlenspiel, K. and Dylla, N. (1993) Experimental investigation of designers' thinking methods and design procedures, *Journal of Engineering Design*, 4(3), 201–212.

Eissen, K. and Steur, R. (2008) *Sketching: Drawing Techniques for Product Designers*. Singapore: Bis Publishers/Page One Publishing.

Evans, M. (1992) Model or prototype which, when and why?, *IDATER 1992 Conference*, Loughborough University (Design and Technology) Loughborough, UK.

Fielden, G.B.R. (1963) *Engineering Design*. London: HMSO.

Fish, J.C. (1996) How sketches work – a cognitive theory for improved system design, Ph.D. Thesis. Loughborough University of Technology.

Fiske, J. (1998) *Kommunikationsteorier: En Introduktion, Wahlström och Widstrand*. Centraltryckeriet: Borås.

Flurscheim, C.H. (1983) *Industrial Design in Engineering*. London: The Design Council.

Goel, V. (1995) *Sketches of Thought*. Cambridge, MA: MIT Press.

Goldschmidt, G. (1997) Capturing indeterminism: representation in the design problem space, *Design Studies*, 18(4), 441–455.

Griffin, A. and Hauser, J.R. (1996) Integrating R&D and marketing: a review and analysis of the literature, *Journal of Product Innovation Management*, 13, 191–215.

Gutwin, C. and Greenberg, S. (1996) Workspace awareness for groupware, in *Proceedings of the Conference on Human Factors in Computing Systems, Vancouver*, pp. 208–209.

Hurst, K. (1999) *Engineering Design Principles*. New York: Arnold Publishers.

IDSA. (2006) *Industrial Designers Society of America (IDSA): About Industrial Design*. http://www.idsa.org/webmodules/articles/anmviewer.asp?a=89&z=23, accessed 2 February 2006.

Jassawalla, A.R. and Sashittal, H.C. (1998) An examination of collaboration in high-technology new product development processes, *Journal of Product Innovation Management*, 15(3), 237–254.

Johnson, S. (1998) What's in a representation, why do we care, and what does it mean? Examining evidence from psychology, *Automation in Construction*, 8(1), 15–24.

Kahn, K.B. (1996) Interdepartmental integration: a definition with implications for product development performance, *Journal of Product Innovation Management*, 13, 137–151.

Kavakli, M., Stephen, A.R. and Ball, L.J. (1998) Structure in idea sketching behaviour, *Design Studies*, 19, 485–517.

Kim, Y.S. and Philpott, M. (2006) *Interdisciplinary Research: Integrated Engineering and Industrial Design*, http://www.engr.uiuc.edu/communications/engineering_research/ 1996/gen1/gen1-9.html, accessed 16 May 2006.

Larkin, J.H. and Simon, H.A. (1987) Why a diagram is (sometimes) worth ten thousand words, *Cognitive Science Journal*, 11, 65–99.

Leenders, M.A.A.M. and Wierenga, B. (2002) The effectiveness of different mechanisms for integrating marketing and R&D, *Journal of Product Innovation Management*, 19, 305–309.

Leonard-Barton, D. (1991) Inanimate integrators: a block of wood speaks, *Design Management Journal*, Summer, 61–67.

Mathew, B.S.J. (1997) Are traditional management tools sufficient for diverse teams?, *Team Performance Management*, 3(1), 3–11.

McGown, A., Green, G. and Rodgers, P.A. (1998) Visible ideas: information patterns of conceptual sketch activity, *Design Studies*, 19(4), 431–453.

Miles, M.B. and Huberman, A.M. (1994) *Qualitative Data Analysis*, 2nd edn. Thousand Oaks, CA: Sage Publications.

Olofsson, E. and Sjölén, K. (2005) *Design Sketching*. Sundsvall, Sweden: Keeos Design Books AB.

Otto, K. and Wood, K. (2001) *Product Design – Techniques in Reverse Engineering and New Product Development*. Englewood Cliffs, NJ: Prentice Hall.

Palmer, S.E. (1987) Fundamental aspects of cognitive representation, in E. Roch and B.B. Lloyds (eds), *Cognition and Categorization*. Hillsdale, NJ: Lawrence Erlbaum Associates.

Pavel, N. (2005) *The Industrial Designer's Guide to Sketching*. Trondheim: Tapir Academic Press.

Pei, E. (2009) Building a common language of design representations for industrial designers and engineering designers, Ph.D. Thesis, Department of Design and Technology, Loughborough University, UK.

Pei, E., Campbell, R.I. and Evans, M. (2010) Development of a tool for building shared representations among industrial designers and engineering designers, *CoDesign Journal*, 6(3), 139–166.

Pei, E., Campbell, R.I. and Evans, M. (2011) A taxonomic classification of visual design representations used by industrial designers and engineering designers, *The Design Journal*, 14(1), 64–91.

Persson, S. (2002) Industrial design and engineering design interaction: studies of influencing factors in Swedish product developing industry. Licentiate Thesis. Department of Product and Production Development, Göteborg: Chalmers University of Technology.

Persson, S. (2005) *Toward Enhanced Interaction between Engineering Design and Industrial Design*. Goteborg, Sweden: Chalmers University of Technology.

Persson, S. and Warell, A. (2003) Relational modes between industrial design and engineering design – a conceptual model for interdisciplinary design work, in Proceedings of the Sixth Asian Design International Conference (ADC'03), Tsukuba,

Pipes, A. (2007) *Drawing for Designers*. London: Laurence King Publishing.

Purcell, T. and Gero, J.S. (1996) Design and other types of fixation, *Design Studies*, 17(4), 363–383.

Rodriguez, W. (1992) *The Modelling of Design Ideas – Graphics and Visualization Techniques for Engineers*. Singapore: McGraw-Hill Book Company.

Rothwell, R. (1992) Successful industrial innovation: critical factors for the 1990s, *R&D Management*, 22(3), 221–239.

Saddler, H.J. (2001) Understanding design representations, *Interactions*, July–August, pp 17–24.

Scrivner, S.A.R., Ball, L.J. and Woodcock, A. (2000) *Collaborative Design – Proceedings of Co-Designing 2000*. London: Springer-Verlag.

Shaw, V. and Shaw, C.T. (1998) Conflict between engineers and marketers: the engineer's perspective, *Industrial Marketing Management*, 27(4), 279–291.

Sproull, L. and Kiesler, S. (1991) Supporting collaborative design groups as design communities, *Design Studies*, 21(2), 187–204.

Stacey, M. and Eckert, C. (2003) Against ambiguity. *Computer Supported Cooperative Work*, 12(2), 153–183.

Strauss, A. and Corbin, J. (1990) *Basics of Qualitative Research: Grounded Theory Procedures and Techniques*. London: Sage Publications.

Suwa, M., Purcell, T. and Gero, J.S. (1998) Macroscopic analysis of design processes based on a scheme for coding designer's cognitive actions, *Design Studies*, 19(4), 455–483.

Tang, J.C. (1991) Findings from observational studies of collaborative work, *International Journal of Man-Machine Studies*, 34, 143–160.

Tjalve, E., Andreasen, M.M. and Frackmann Schmidt, F. (1979) *Engineering Graphic Modelling – A Workbook for Design Engineers*. London: Butterworth & Co.

Tovey, M. (1989) Drawing and CAD in industrial design, *Design Studies*, 10(1), 24–39.

Tovey, M. (1994) Form creation techniques for automotive CAD, *Design Studies* 15 (1), 85–114.

Tovey, M. (1997) Styling and design: intuition and analysis in industrial design, *Design Studies*, 18(1), 5–31.

Toye, G., Cutkosky, M.R., Leifer, L.J., Tenenbaum, M. and Glicksman, J. (1993) SHARE: a methodology and environment for collaborative product development, in *Post Proceedings of the IEEE Infrastructure for Collaborative Enterprises*, pp. 1–16.

Ullman, D.G., Dietterich, T.G. and Stauffer, L.A. (1988) A model of the mechanical design process based on empirical data, *Artificial Intelligence for Engineering Design, Analysis and Manufacturing*, 2(1), 33–52.

Ulrich, K.T. and Eppinger, S.D. (2000) *Product Design and Development*. Boston, MA: McGraw-Hill Inc.

Veveris, M. (1994) The importance of the use of physical engineering models in design. IDATER 1994 Conference: Loughborough University (Design and Technology).

Chapter 10

The Use of Design Case Studies in Design Education

SEYMOUR ROWORTH-STOKES AND TIM BALL

Introduction

Case studies for teaching purposes are commonplace in many disciplines such as management, law, health and life sciences. They are often used to demonstrate principles, precedents and insight into 'real-life' contemporary phenomena (the here and now) set against critical incidents, happenings or events over time (cause and effect). Yet there appears to be a dearth of literature surrounding the use and role of case studies in design.

This chapter considers the role and use of case studies in design education following a systematic audit of 223 'cases' in the design research literature by type, subject and field of research. The findings suggest that there are multiple descriptions and interpretations of the term 'case study'. Although there are more traditional approaches to case-based reasoning in architecture, overall a more innovative and dynamic discourse appears to be emerging. This is explored further with an example of innovative pedagogy being employed within a studio-based design-learning environment which develops the concept of case-based designing.

Background

Case studies are increasingly being used to illustrate, demonstrate and provide evidence for issues in design research.(Roworth Stokes 2006) The method has been employed across many areas including new product development, product innovation, design behaviour, risk evaluation, and supply chain management (see for example Bussracumpakorn, 2002; Horne-Martin et al.,

2002; Cooper et al., 2002) Case studies are often used to investigate phenomena in 'real-life' situations, where we want to understand factors surrounding the design process. Yin (1993) defines case studies as being appropriate when contemporary phenomena are to be investigated in their real-life context; when the boundaries of the phenomena and the context are blurred; and multiple sources of evidence are used. Chetty (1996) argues that its main strength is its ability to measure and record behaviour and that multiple sources of data can be brought together to gain as full an insight as possible.

> *These include documentation, archival records, interviews, direct observation, participant-observation and physical artefacts. (Chetty, 1996: 74)*

Yet case study has been seen as being 'soft' due to the difficulty of making generalisations from a site-specific context and the common journalistic style of reporting a single 'case' as being typical of a wider phenomenon (Yin, 1993). This can be further complicated in multiple case study research when a massive amount of data is generated with limited structure to make sense of it. This has sometimes led exponents such as Yin to perceive its value as being under appreciated:

> *Most people use it as a method of last resort, and even then they use it with uneasiness and uncertainty. Despite the availability of key works on how to do case study research. (Yin, 1993: 40)*

Another problem is that case study method is a broad term and encompasses many approaches. It is a hybrid, even though it inevitably errs on the side of qualitative research, due to the need to understand 'how' and 'why' questions. Langrish (1993) rightly points out that these perspectives originate from a different worldview with the 'physics' approach on the one hand, which looks for underlying principles, and the 'biological' approach on the other, which glorifies diversity.

> *There are two different traditions in research and that they can be labelled the physics approach and the biological approach. Case studies are in the latter domain and are therefore not understood by people whose world view belongs to the physics domain. (Langrish, 1993: 357)*

Both Breslin and Buchannan (2008) as well as Langrish (1993) highlight the important relationship and role of case studies for teaching alongside the construction of theory through research.

Case studies have a rich history for exploring the space between the world of theory and the experience of practice. It is one thing to have an idea and another thing to make that idea concrete and real. Designers, by the nature of what they do, must become skilled at moving between those two places. But recognizing and understanding the transition from the one place to the other, and back again, is difficult. Case studies are a useful tool for research and teaching that focus on the transition between theory and practice. The format has been widely used in other disciplines, and it can be used effectively in design. (Breslin and Buchannan, 2008: 36)

This also brings into focus the more practical problem of creating a detailed and transparent process of interpretation and analysis to ensure the relationship between evidence, concept development and theory remains valid and rigorous.

However, nearly 20 years after Langrish's first clarification on the subject and a more recently call by Breslin and Buchanann for 'fourth-order design' (2008: 40) to extend theory and the study of practice through case studies, there still remains a significant gap in our knowledge of just what case study material is out there.

Methodology

This chapter seeks to explore the use of case study method in design education. We were particularly interested in the way case studies had been utilised to derive theory, and to propose constructs and/or principles in an empirical sense. This would involve a comprehensive literature review to fully appreciate the methodology employed and to eradicate any common misunderstanding concerning use of the terms 'case/s' and 'study/ies' by any open and non-corroborated online search.

First, leading design research journals were identified. Mindful of the emerging nature of the field of design research since the 1960s, we were aware of the need to understand the range and profile of contributions within the field. Cross (2006) suggests that design research 'came of age' in the 1980s following the founding of the first journal for design research in 1979, *Design Studies*.

The number of journals has continued to grow in subsequent years and Sugiyama (2003) recognised the relationship between a growing number of doctoral programmes in design and a growing design research community. In 2008, Friedman et al. (2008) identified 173 journal titles representing the diversity of publications and routes to dissemination in the field. A comprehensive review of all these journals is beyond the scope of this chapter, but the survey of academics undertaken by Swinburne University and RMIT (Royal Melbourne Institute of Technology) to rank journals is a useful starting point to prioritise publications. This suggests the four leading journals are (Table 10.1).

Table 10.1   The journals by rank order, data originally cited in Friedman et al. (2008, Annex A)

The journals by rank order	Cited
Design Studies	152
Design Issues	146
International Journal of Design	85
Design Journal	84

It was decided to undertake a preliminary search using the qualifier 'case study' to determine articles for further evaluation within this group of publications. Each of the papers was then read to eradicate any erroneous classification. The definition of a case study cited by Yin (1993) and Chetty (1996) was used to inform this process (Table 10.2).

Table 10.2   Criteria to define a 'case study' drawn from Yin (1993) and Chetty (1996)

Indicators	Evidence
Phenomena is investigated in a 'real-life' context	Background material is provided
The boundaries of the phenomena and the context are blurred	The 'case' is specified and described (e.g. project, person, product)
Multiple sources of evidence are used to explore and understand outcomes to gain as full an insight as possible	Such as documentation, archival records, interviews, direct observation, participant-observation and physical artefacts

This yielded the results shown in Table 10.3.

Table 10.3   **Search results**

	Total number of articles searched	**'Case study' qualifier**	**Total (meeting definition)**	**Percentage**
Design Studies	1,726 (since 1979)	133	121	7
Design Issues	539	75	48	9
The Design Journal	307 (since 1997)	34	29	9.5
International Journal of Design	79 (since 2007)	25	25	31.5

In total 44 papers were removed at this stage. The reasons included:

- Citing findings from previous studies based on case studies in literature reviews

- The words 'case' and 'study' being used independently for differing purposes

- Reference to 'case study' in the keywords which was did not meet the definition above.

Design Issues originally published just two editions a year which may account for the smaller total and the *International Journal of Design*, which has a proportionately higher number than any other journal, was only launched in 2007. Interestingly this latter journal has a dedicated 'Design Case Studies' section. However, *Design Studies* has the longest publication history (1979) and contains the largest number of case studies, greater than the other three combined. As the leading design research journal it forms the basis of the analysis that follows.

Analysis

Case study methodology has many variations and indeed opposing viewpoints. This is exemplified by the debate surrounding the promotion of the exploratory/ intrinsic case or classic case, as referred to by Dyer and Wilkins (1991), and the multiple case study approach proposed by Eisenhardt (1989).

The essence of case study research is the careful study of a <u>single</u> case that leads researchers to see new theoretical relationships and to question old ones. (Dyer and Wilkins, 1991: 614)

While Dyer and Wilkins believe in the deep understanding of a single rich 'story', Yin (1993) supports the view that case study research can be used to test hypotheses in a deductive manner by deriving a sample of cases that are 'explanatory' in nature. On the other hand, Eisenhardt (1989) sees the method as being more appropriate to build theory in an inductive manner. However, both Yin and Eisenhardt agree that the method is capable of developing generalisations through strict adherence to a methodological framework – '*case studies, like experiments, are generalisable to theoretical propositions*' (Yin, 1984: 21).

To develop an appropriate categorisation of the cited case studies it was important to determine features to explicate some of the common forms of the method. Table 10.4 was used to guide the evaluation of each article.

Table 10.4   Different approaches to case study research

Type	Description	Methodological approach	Ontological/ epistemological implications	References
Exploratory/ intrinsic/classic case	Used where there are signs of limited knowledge To 'explore the territory' 'What, 'how', 'where', 'when' research questions To gain a better 'deep' understanding Illustrates a particular trait Explore abstract concept or phenomena, 'the one-off' 'How' or 'why' research questions	Develop theory and then test where possible In depth using range of methods and observation over time Empathy essential to building trust with respondents	Subjective Can be ethnographic – transformative and empowering or part of multiple case approach Largely inductive and qualitative Can illustrate existing argument or predisposition – constructivist/ ideologist	Yin (1993) Stake (1994) Dyer and Wilkins (1991)

Table 10.4   *Concluded*

Explanatory/ instrumental	Test cause and effect relationship Insight into an issue or refinement of a theory Case chosen as part of larger research interest	Used to test theory Large range of research methods	Hypotheses testing Errs toward deductive Can lead to theory-building Objective Realist	Yin (1993) Stake (1994)
Collective/ multiple case	Instrumental study in multiple Cases chosen due to theoretical representation of phenomena, population or general condition	As above, and Allows cross-case comparison Usually between 4 to 10 cases in practice Develop theory Methodological 'framework' essential Can be inductive or deductive Likely to lead to theory-building and generalisations	More post positivist than phenomenological Objective Realist	Stake (1994) and Eisenhardt (1989)
Common/ features	Focused on site-specific instance/s – 'real life' Ability to understand complex interaction of phenomena in play – 'how' and 'why' questions	Multiple sources of evidence are used Boundaries between phenomenon and context appear blurred	Nearly all ontological/ epistemological positions are possible	

Findings

Figure 10.1 provides an overview of the split between the three types of case study indentified. There were twice as many Exploratory/Intrinsic case studies as Explanatory/Instrumental (50 compared to 25) with more than a third of the total being Collective/Multiple case studies.

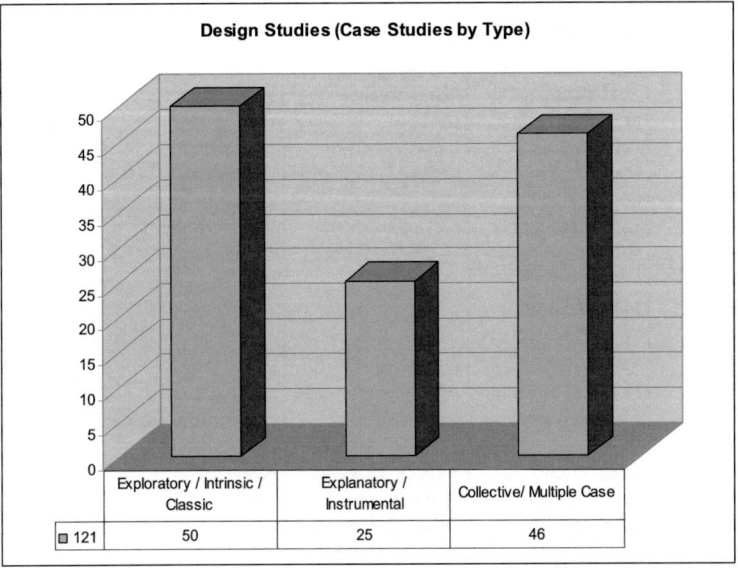

Figure 10.1   Design studies by case study by type

Each of the cases was then classified by subject type. This process was conducted using an 'emic' approach as described by Spradley and McCurdy (1979) whereby the text was used to describe the subject definition in as near to the authors own interpretation as possible.

> *Emic descriptions of sound depended on discovering the native categories and perceptions. In the same way, etic descriptions of behaviour, on the other hand, of sound or anything else are based on categories created by the investigator, and are usually employed to compare things cross-culturally. (Spradley and McCurdy, 1979: 231)*

Figure 10.2 shows the results of this classification.

Figure 10.2 indicates there is a wide range of design disciplines covered but there is a significant variation in the representation of 'cases'. The 'Architecture/ built environment' and 'Product design' cases represent nearly half of the total number (57 combined out of 121) with only a few cases evident in areas such as 'Contemporary crafts/textiles' and 'Graphic design'. It is worth noting that these findings seem disproportionate to the number of students studying in design-related areas at undergraduate level when one would expect areas such as fashion and graphic design to be much better represented.

Case studies by theme, indicative of fields of inquiry in design research, are now presented in Figure 10.3.

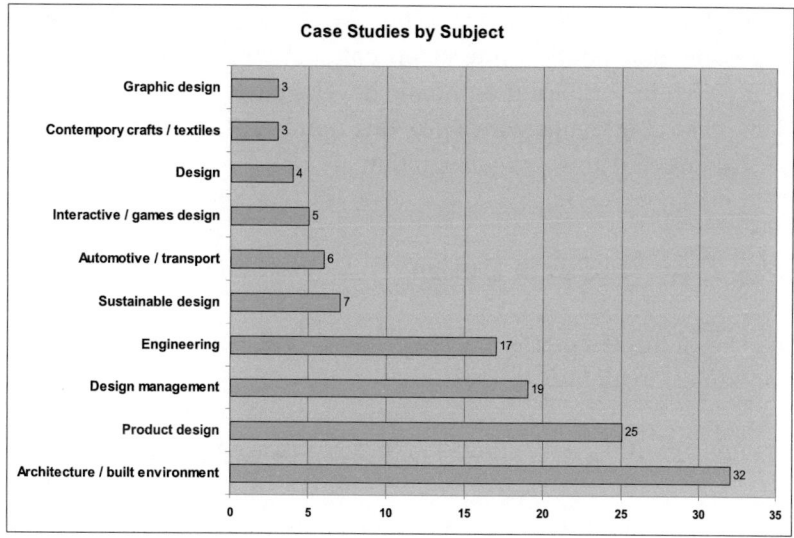

Figure 10.2   Case studies by subject

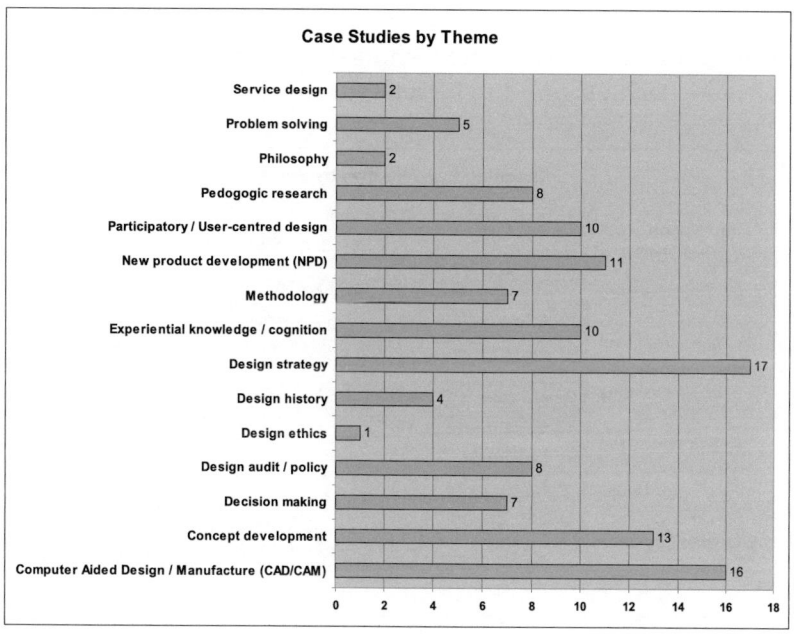

Figure 10.3   Case studies by theme

As with the subject coverage, while there is a wide-ranging breadth of areas represented there are some significant areas associated with innovation, such as 'NPD (New Product Development)', 'Design Strategy' and 'Concept Development' alongside the processes associated with integrated rapid-manufacturing practice through 'CAD/CAM'. To some extent, this reflects the growing social and political interest in national innovation systems and the role of research in furthering economic development. It is also interesting to note some emerging trends with a growing number of cases in the burgeoning field of 'Experiential knowledge/cognition'.

Teaching Case Studies in Design

As the focus of this chapter is teaching related case studies a separate analysis was undertaken to identify those studies that:

- sought to utilise case studies to further pedagogic research;

- identified implications for teaching and learning; and/or

- located and placed the context for the case within the educational environment (often involving students on case-based projects).

Twenty teaching-based cases were identified out of the original 121 and these are now presented in Figure 10.4 by subject and theme.

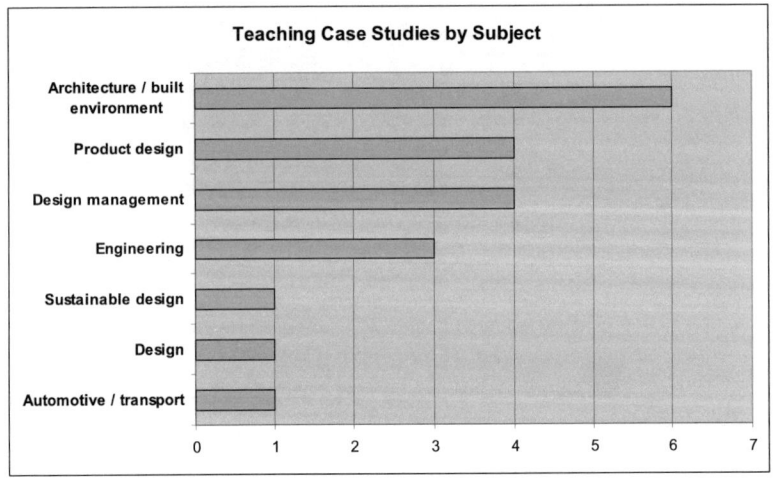

Figure 10.4   Teaching case studies by subject

Unsurprisingly the subject coverage of this subset has many similarities with the overall population of case studies. Architecture and the built environment still dominate the cases identified. This may suggest a stronger case-based teaching pedagogy exists in this subject area. Akin (2002) has suggested in a study of case-based instruction strategies in architecture that cases are often recalled in architectural education to illustrate a particular design issue. For example the functional aspects of a building, how a design constraint was overcome and for evaluative purposes to weigh up how one design solution may have advantages over another. It is also suggested by Akin (2002) that case-based reasoning is rooted in the studio. This team-based professional practice model documents and records each building as a 'case' which can be observed, analysed and understood in its own context. This may also account for the proportionately higher number of product design and engineering cases as they tend to form part of the credentialisation process for design teams when promoting and communicating the value of the services they offer. These can be grouped by theme as in Figure 10.5.

Three main types are evident in Figure 10.6. The first is associated with methodology, with a specific focus on new or innovative methods in learning and teaching practice and often situated directly in the learning environment. Second, Practice/Case-based Projects investigate and 'frame' findings of problem-based projects and consider implications for teaching practice.Finally, Experimental/Comparative cases are those that compare and contrast specific characteristics or variables in design practice by evaluating the findings of projects cross case.

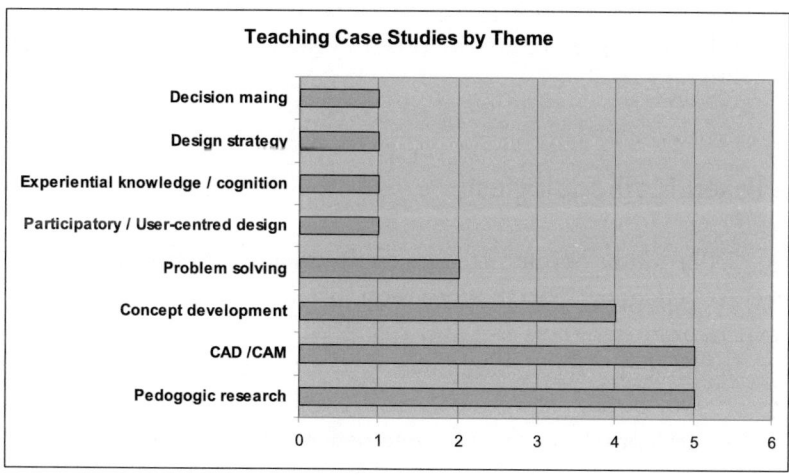

Figure 10.5   **Teaching case studies by theme**

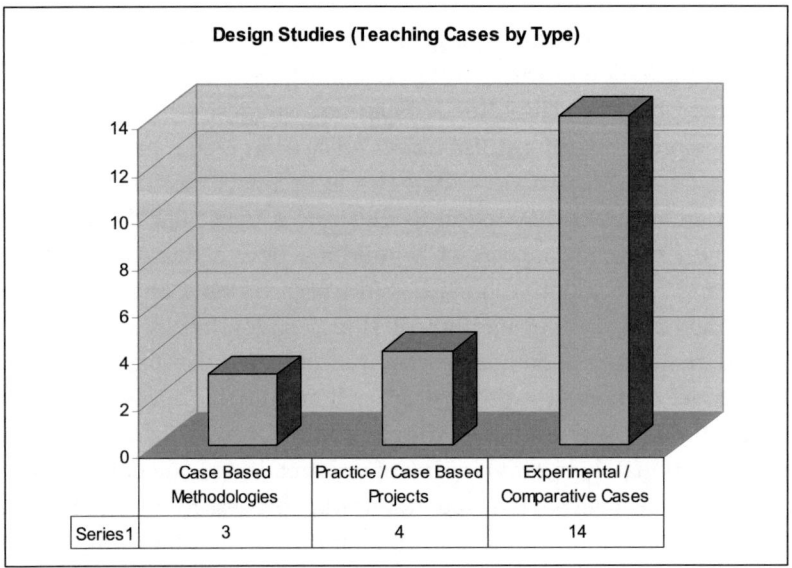

Figure 10.6   Teaching cases by type

It should also be noted that Chiu's (2002) study of design communication through collaboration cross two types here, the Experimental/Comparative Cases and Practice/Case-based Projects. The paper considers the results of case studies in architectural practice and design studios and develops a process model of design collaboration. The findings suggests that the success of a design project can be affected by a structured organisation and computer-supported collaborative work requires key tasks to be communicated to individuals, groups and those more broadly involved in the project. These areas are now described briefly.

Case Based Methodologies

Akin's (2002) study reinforces the tradition of case-based reasoning in architecture pedagogy. The study was conducted in order to better understand how experienced designers use and reference cases in their practice as part of the development of a database developed to support studio-based teaching instruction. Out of 55 instances of visual/graphic references being identified within the conceptual design, 31 contained references to cases of which 25 were unique. Examples ranged from iconic modernist buildings such as the Guggenheim by Frank Lloyd Wright and the Centre Georges Pompidou by

Piano and Rogers as well as some less popular and more local historic references such as the Pittsburgh Athletic Association in Pennsylvania building with a clearly Venetian Renaissance influence designed by Benno Janssen. Akin's (2002) research found that while cases were commonly used in architectural design there was a dearth of research into the use of such cases. The database itself has the potential to enhance the ability to organise and make more accessible material surrounding regularly cited cases, however it is notable that they found it difficult to generate full and complete accounts of many of the older designs listed.

Burdick and Willis (2011) have undertaken an analysis of the literature on education to explore the role of design thinking in digital learning and digital scholarship. They develop their argument through the use of cases which recognise the value of design thinking as being situated, interpretive, and user-oriented. They draw upon examples such as the Institute of Play in New York City which has used abductive reasoning originally considered by Nigel Cross as a means to promote innovation, discovery, strategic thinking and play within the curriculum.

Oxman and Oxman (1992) develop an indexed 'case base' from which to explore cognitive design precedents in terms of the issues faced by architectural designers, as they develop concept and form. Two specific concepts are expanded upon around refinement and adaption.

Practice/Case-based Methodologies

The Practice/Case-based Methodologies are again dominated by architectural case studies. Baynes (1982) case study explores the study of buildings and environments as part of the art curriculum in secondary schools. It raised questions about the development of design awareness in children, and the social relevance of expressing and communicating ideas.

Lee (2009) has created a typology of project-based learning from a study of 1,500 students studying design subjects resulting in 6 project types which cover degrees of complexity and student autonomy, and learner skill/knowledge and problem difficulty. This range of variables and overlap between project types is described by Lee as raising significant challenges for lecturers and students when trying to maintain a high level of understanding in a dynamic and rapidly changing learning environment.

Bilda, Gero and Purcell (2006) analysed three architecture practice case studies to identify whether sketching was essential for conceptual designing in order to determine whether it was required in design education. They confirmed that it did.

Experimental/Comparative Case

Goldschmidt and Tatsa (2005) present two illustrative case studies within an architectural school studio to determine the role and influence of ideas. They conclude that ideas are important but the role of the 'critical' idea, to challenge convention or commonly held beliefs or conventions, is equally as important as the conceptual idea leading to a final approval.

Sachs (1999) refers to two cases, a studio of students in an BArch architecture programme in Israel and another involving Bachelor of Art students in the United States of America. Through comparative appraisal by observation, discussion and interview, the issue of 'stuckness' is explored. The studio space is proposed as a means to resolving these situations as an open and reflective space whereupon combined knowledge and expertise can be brought to bear to make sense of, and indeed resolve, problems that at first appear insurmountable.

In a controlled study, Christiaans and van Andel (1993) compared approaches to presenting students in industrial design around the psychological aspects of design. They observed that the group who had received a more detailed briefing generated better features and characteristics in their design solutions that met user expectations.

Jonson (2005) undertook a comparison of conceptual tools between education and professional practice with five cases presented on undergraduate design students design work with an equivalent number of design practitioners undertaking the same project. Verbalisation, rather than freehand sketching, was the major conceptual tool for getting started although the findings suggest practitioners rely on words more than students while universally, more than four times as many breakthroughs were generated through discussion and verbal engagement than for sketching and computer modelling.

An Illustrative Case Study

In order to explore these issues in context, an illustrative case study is now presented. The aim here is to consider the use, role and interpretation of case-based teaching in a primarily studio-based learning environment. The case is a module based within the Industrial Design Department at Coventry University, home to one of the countries largest range of studio-based design courses for undergraduate students.

The module involves a cohort of 60 in self-selecting groups of 3 and runs from January to early June, at level 2. It seeks to promote and consolidate a broad range of skills and knowledge while furthering critical judgment. It uses an innovative approach to case-based design pedagogy by challenging students to place their course work in the context of a diverse range of contemporary and historical design practice through the use of case studies. The overarching theme linking the various projects and the delivery of activities throughout the module is the word 'EAT'. Not intended to be taken literally, but reflecting a fair comparison with some aspects of building confidence in student designers and introducing the enriching role of metaphors, allegories and analogous thinking to everyday design activity.

The module commences with the whole cohort being introduced to a series of 10 case studies of contemporary architects work. Each group then takes one architect from the list below, as the starting point for detailed research surrounding the philosophy and approach to their work and practice:

- Frank Gehry

- Zaha Hadid

- Frank Lloyd Wright

- Ludwig Mies van der Rohe

- Renzo Piano

- Jean Nouvel

- Le Corbusier

- Santiago Calatrava

- Ron Arad

- Alva Aalto

A rough outline and schedule is given in Table 10.5.

Table 10.5   Teaching schedule for module

Phase	Method/approach	Aim/objective	Learning outcomes
Research	Group-based activity (each draws 1 lot to determine which architects are selected) Collective learning and contextualisation of all 10 architects (analogy of literature review drawn from individual contributions)	Compile a presentation made to the rest of the cohort (using mood/ theme boards) Each group researches the architect's work and presents it as a synopsis to the cohort	To develop succinct, accurate, referenced visual and verbal communication skills with confidence To investigate the architect's life, work and design output, alongside their training, professional development and design philosophies To undertake and demonstrate research, analysis and critical thinking To jointly secure an understanding of the entire field being reviewed
Concept	Studio-based project	Generate proposals for a set of (min. 3) eating implements) + 3 eating/drinking vessels or other objects, e.g. serving implements or other aspects of the dining experience	Prepare a written design brief and specification Demonstrate the ability to work collaboratively To develop a set of tableware that both complements and expresses characteristics of the architect's work and philosophy
Final design proposals	Design development	Form models and supporting drawings – recorded for inclusion as pages in portfolios Compile and present models, drawings and a PowerPoint presentation (max. 10 slides)	To demonstrate an active engagement in the group design process and proposals (each group member must make a verbal contribution to the presentation) Generate innovative design concepts and express these through 2D concept sketches and translate these into 3D form, deliver a formal presentation of a final design solution to peers and staff supported by evidence of research, analysis, and design concept development To consider the nature of a hierarchy of design activity and to reflect on relationships product design has within the built environment, materially, emotionally and stylistically in addition to how the work of other designers influences and shapes creative thinking

The research phase allows for a dense amount of case and contextual studies to be covered as an integrated element of the module, which in turn informed a linked design task. Design proposals are developed through sketch drawing and modelling and differing degrees of engagement with manufacturing of prototypes supported by university workshops, using wood, plastics, metals and ceramics. Digital and Rapid Protyping machines were also used to produce facsimile items, moulds and other tooling for use in secondary processes.

Importantly, the module confronts design students' ability to appraise and analyse existing design approaches and philosophies before embarking on a design process in the guise of the selected architect. To successfully achieve the learning outcomes, a balance must be struck between the desire to be novel and innovative and an ability to work collectively in a group guided by a commonly understood and integrated design philosophy.

The Evolution of Case-based Designing (CBD)

Figure 10.7 is from the students' presentation (which served as a concise visual summary and accompanying verbal narrative of the project); this slide identified the key characteristics the group identified in Aalto's work.

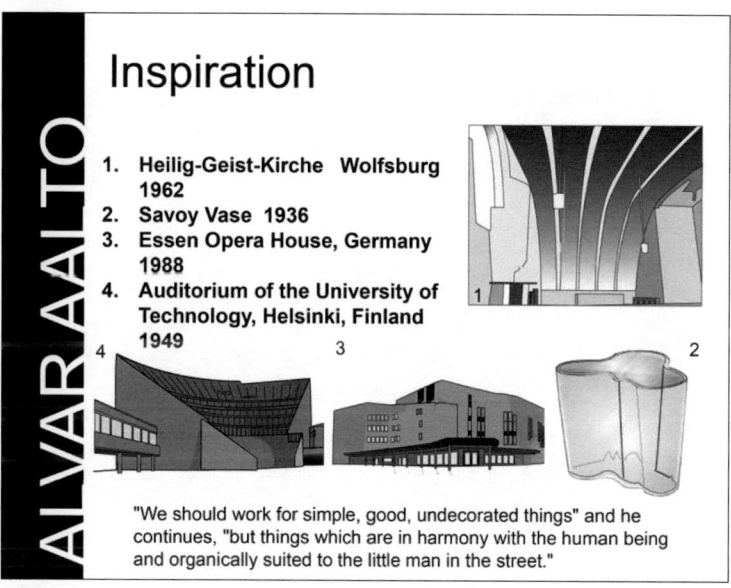

Figure 10.7   Research board

The presentation board uses a short chronological history of Aalto's major work as the basis to explore and define line, materials and proportion. Cultural, philosophical and contemporary technological developments in Aalto's practice were described alongside examples of tableware proposals that would complement Aalto's buildings. The students expressed this from the perspective of his vision of the built environment.

Figure 10.8 is a composite image; simple line drawings have been used to explore initial concepts, echoing both positives and negatives of the referenced buildings and products. The use of simple use of shadows adds some sense of solidity, but also suggests an architectural, totemic idiom. While it's unclear as to whether this was fully intended, it's a quality that might not have existed had the students referenced product-scale objects alone, and reinforces the value in looking for inspiration outside the immediate domain in question.

The brief was open and invited concepts for eating implements, rather than cutlery per se, this was intended to require the students to explore the tasks of eating in parallel with their visual, formal, ergonomic, material and process inquiries. An integrative approach is evident in the development of these proposals. Although the proposals, as they become more developed and refined (reflecting a range of techniques beyond line drawings), tend to conform to expected design development norms, they can also be seen to maintain a distillation of the characteristics they aspired to – essentially the mantra of the architect they have studied.

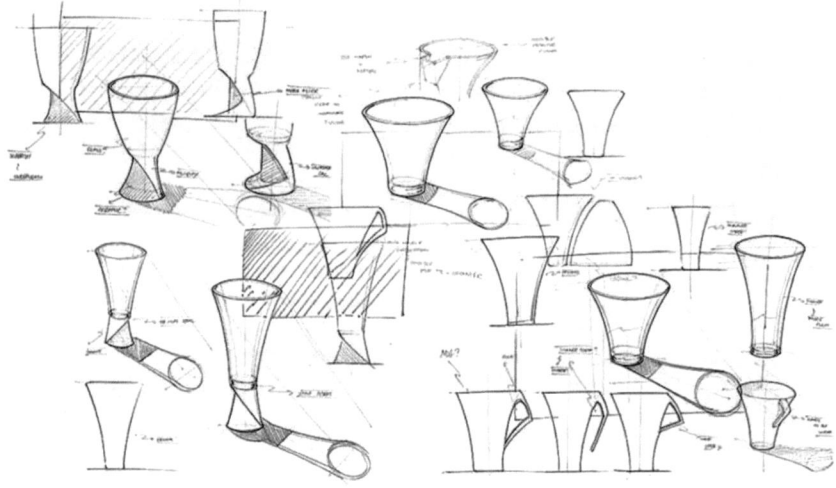

Figure 10.8   Concept sketches (1)

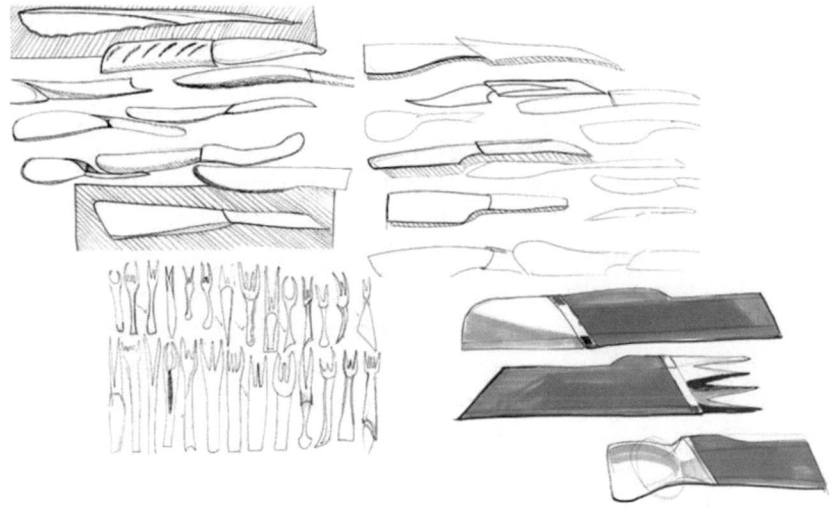

Figure 10.9   Concept sketches (2)

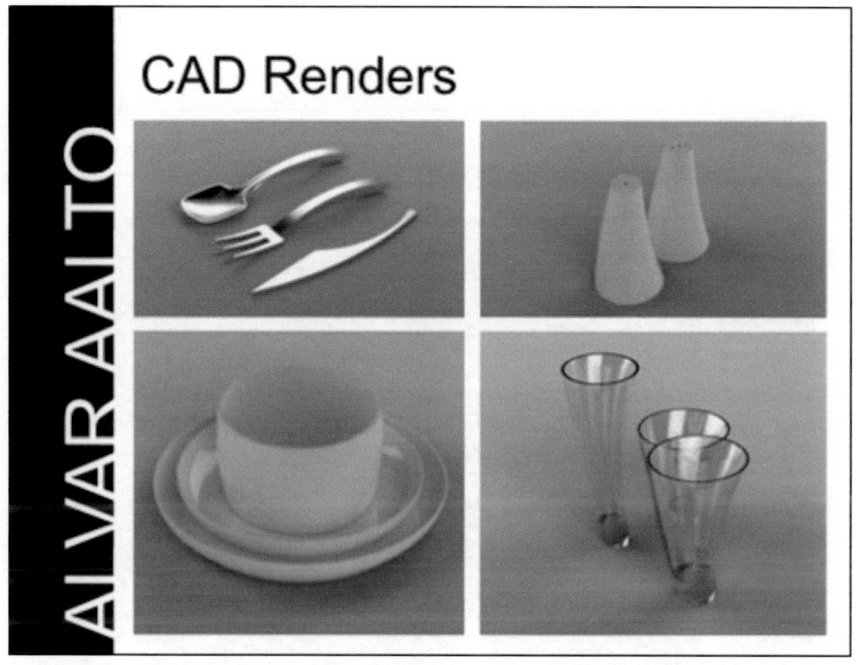

Figure 10.10   Final design proposals (1)

Source: By permission of Coventry University.

While these high-resolution CAD images might appear to lack connection with the physical reality of materials and manufacturing processes, Figure 10.11 shows a sample of extensive and sensitive testing of forms through prototypes generated from turned blue foam as slip casting tools for ceramic and reinforced plastic products. The group has engagement with both human and food scale, and an awareness of product juxtaposition with interior features of built environments.

Underpinning the whole module (process) is the concept that learning from case studies provides multiple, informed starting points for design development. The project demonstrates an iterative, integrative learning process that began with contextual study of the adjacent creative field of architecture and concluded with new, tangible product proposals. Using an architect as a 'case' fosters curiosity and independence in students' ability to identify relevant problems and make meaningful design interventions. Students are naturally introduced to a rich heritage of historical and contemporary design activity, combined with a range of evolving techniques for developing, expressing and communicating new design intentions. In summary, the module fosters an understanding of theory through practice and promotes sustained independent (and interdependent) study.

Figure 10.11   Final design proposals (2)

Source: By permission of Coventry University.

Architects have often designed tableware as in the instance of Arne Jacobsen's St Catherine's College (opened 1962) with his furniture, cutlery and lighting. This is well before the recent trend for manufacturer commissioned projects such as *Tonale* by Alessi designed by David Chipperfield (2009). Such instances are cited during lectures supporting the module and used to question the micro versus macro approaches that inform both architectural and product design scale and development processes.

Design education frequently operates outside conventional learning frameworks when it comes to the acquisition of skill and knowledge. While there might be specific, desirable learning outcomes, the journey towards achieving them is much harder to define given that, by its very nature, the intention is to produce unexpected outcomes. On this point, it is acknowledged much design theory and practice relates to the development of visual judgement. While other subjects might deal with reasoning linked to objective themes, design students are required to address the additional, subjective values and the uncertainties arising from their need to establish and codify aspects of visual literacy. There is benefit to be derived from creating awareness of reference points that are related, but differing in scale and or function. By examining architectural scale concepts (and realities), product design students can be seen to achieve a greater degree of engagement than that derived from emulating product design case studies alone, alongside a richer awareness of problem-solving, historical movements and international perspective.

Therefore, the discourse arising from a project – the process – is often of equal if not greater value than the outcome. This is observed by Mau in his wry yet perceptive *Incomplete Manifesto for Growth*:

> *Process is more important than outcome. When the outcome drives the process we will only ever go to where we've already been. If process drives the outcome we may not know where we're going, but we will know we want to be there. (Mau, 1998: 1)*

For example, a simple analogy for the 'live project' that aims to reflect commercial reality in an educational environment, would be throwing a non-swimmer in at the deep end. Of course, the swimming pool, with known edges, is supervised and contains little or no risk – excepting perhaps personal pride- a situation that varies from the murky, open waters of commercial practice of sink or swim. As student progression approaches professional engagement, the imperative becomes the move away from theory through practice towards greater control over creative and commercial certainty as practice through theory. A response from one student reflects this complex and interrelated exploration of skills and knowledge:

> *I have really improved my skills in research. In previous projects I feel that I wouldn't have gone into as much detail on the materials and the processes ... I had a better knowledge of the product and the materials and that really helped me get to a good concept product.*

Unlike case-based reasoning and case-based instruction approaches, which mirror traditional teaching case delivery to understand good or bad designs, this example suggests cases can also be used in design pedagogy to challenge the very role of the designer as creator/originator. Here the case acts as a basis for the designer to be a facilitator and interpreter; a collaborator and co-designer.

Feedback and Learning Outcomes

It should be noted that some of the feedback in this section was derived from statements made by students with English as a second language. This is not unusual or representative of a disproportionate number within the cohort; however it is often the case that such students offer comments, phrases and observations of disarming honesty and depth, more so than their native English-speaking peers (though this could in itself, form the basis of another separate paper).

Using Strauss and Corbin's (1990) approach to coding and categorisation, three axial concepts were identified from students' module feedback: (a) *Constant reflection and refinement*; (b) *Using contextual knowledge to inform visual language*; and (c) *Development of 'integrative awareness'*. The concepts are described below using selected quotes which describe the learning experience in respondent's own words.

a)   *Constant reflection and refinement* based upon previous action – embedding the roles of creative thinking, problem identification and critical thinking skills as vital elements, integral to all design activity while judging when to apply appropriate methods; not necessarily following a rigid, linear progression.

> *I learn more about how to make the product suit to customer, for example how to know what are their expectations and difficulties. I need*

to improve knowledge about various kinds of materials. To master the production process, this will be good for my design. Keep drawing and thinking.

This project has been a great test for me in terms of belief and confidence. It has highlighted ... unrealistic expectations of myself to constantly create 'genius' designs ... that to fail is okay, and I will have ideas that won't always be perfect.

The comments above demonstrate the emergence of self/professional awareness, indeed they reinforce the view of Lawson (1997: 11) who describes design as a *'highly complex and sophisticated skill. It is not a mystical ability given only to those with recondite powers but a skill which, for many must be practiced rather like the playing of a sport or a musical instrument'*.

This also relates to an observation by Theroux (1999) offered to the students at the outset of the module when the theme 'EAT' and the analogy drawn between food preparation and design was drawn: *'Cooking requires confident guesswork and improvisation – experimentation and substitution, dealing with failure and uncertainty in a creative way.'*

By way of further contextualising of the delivery and concept for the module:

One of the questions that pop up occasionally in various fora is 'how do you define design?' For a different (light-hearted or low-calorie) way of looking at this, design could be considered a bit like cooking: it's part constructive discontent (it could taste better), part empathy (what do those who will eat the food expect, anticipate), part savvy (you have to know some of the basics to cooking), topped off with a good dose of creativity (seeing beyond the given parameters or the confines of the recipe itself). And cooking is incredibly iterative, just like design; no recipe is sacrosanct – they can be bettered, simplified, adapted to evolving palates, etc. The result may well be failure (too salty, too sweet) but more often than not the result will surprise, thrill and make those who have the pleasure of enjoying it come back for seconds. Constructive discontent, empathy, savvy and creativity – a tasty recipe. Right, off to the kitchen ... (Abaster, 2009)

b)   *Using contextual knowledge to inform visual language* – articulating and communicating perceptual qualities which have meaning for the consumer and offer brand value such as form, colour, texture, (space), material, finish, tone, mass, etc.

Every design problem starts with the same questions, Who, What, Where, When, How, Why? A good product can make consumers' lives more convenient, and a good designer need to know what consumers want.

I have found myself broadening my design horizon by looking to design for a particular clientele in mind. I have enjoyed this project as the research sections have allowed me to fully investigate the various aspects of designing for humans.

I've learnt that not everything has to be complex and the best solution is sometimes the simplest one. My interpretation of visual language and application of it has improved through study of colour, material, finish (CMF).

These accounts can be seen to acknowledge the role of case studies in the development of historical and contemporary practice as well the importance of contextual studies informing personal practice. In some instances, they appear to provide a form 'permission to proceed'; derived from interrogation of a specialist subject, example or requirement and – where appropriate – viewing of precedents that might be improved upon by virtue of a more contemporary appearance; alternatively they promote interpretations of established principles applied to new situations, technologies and/or human behaviour.

c) *Development of 'integrative awareness'* – understanding the need for a breadth and depth of knowledge and expertise covering many sub-related disciplines such as psychology, ergonomics (human factors), manufacturing, marketing and an ability to juggle complex relationships between them and the languages that underpin them.

I have had to carry out a lot of individual primary and secondary research … I need to fully engage with the users and dive straight into user-focused design, through things like personas and ergonomics. One of the things that I need to improve on as a designer is having the confidence to create dialogues and share my thoughts with other designers.

> *I managed to develop better my skills in areas like: working in a team, experimenting with materials, working on CAD and mainly for me something totally new was ergonomics and persona tools ... Behind a spoon there is so much more than a piece of metal.*

> *I have learnt that computers cannot be relied on to save the day with a perfect image of the final product, and nothing beats having models that you can manipulate and feel in your hands to see what can and can't work.*

> *For the first time in my studies, I feel as though I have designed a product and got to know it inside out, down to every single component, and am extremely happy and impressed with myself for designing something that I believe works well enough to pitch. This has made me feel like a designer for the first time, as I have learnt a great deal from this project and really increased my skills and confidence in problem-solving, sketching, sketch modelling, CAD and consulting with other students in a professional manner.*

In their observations of the overall module content and how this might have contributed to their progress, the students offer a wide variety of interpretations of their own developing practice. These are sometimes in relation to a personal journey, but increasingly, in terms of a more worldly awareness. In the context of diverse design practice, they identify the need to juggle a much broader range of considerations, judgements and awareness of other specialists who might help manifest their proposals or impact upon their professional development. In relation to a deeper, more profound engagement with the subject, some articulate their recognition of aspects that impressed them at the outset but how these appeared relatively shallow in relation to the need for the comprehensive, integrative approach the module seeks to impress and deliver. A key expression used frequently throughout the module being the ability '*to design for people who aren't you*'.

It could be suggested that the ability to recognise and conceptualise this 'lens' to frame the design process is the ultimate realisation of all these underpinning attributes. The 'genius' designer observation lends weight to this notion as it suggests a 'detachment from self'; an the ability to take risks while recognising parameters for successful design outcomes are not always associated with the personal motives of an individual designer, one of the key attributes as it tends to reinforce the ability to step outside 'designing for oneself'.

Issues of Rigor and Validity

Breslin and Buchanan (2008) suggest that case studies have the potential to play an important role in understanding and developing theory from practice.

> *The core principles of the discipline are taught through practice, and are presented as part of a solution for a specific problem. For this reason, the learning from one project may not survive in the transition to other projects and problems ... As a result, design case studies have a more difficult, two part job of establishing theory and, at the same time, creating or recreating a bridge back to the practical ... But there is a further opportunity in design case studies, the opportunity to begin talking about theory as theory instead of merely a practical application of wisdom and rules-of-thumb. (Breslin and Buchanan, 2008: 39)*

If design case studies are to be used to further the development of theory in design education it follows that the cases should demonstrate a sound basis for empirical evidence to illustrate practice. We found several very good examples of Yin's (1993) 'Exploratory/Intrinsic', 'Experimental/Instrumental' and 'Multiple' cases. Lehoux et al.'s (2011) multiple case study provides a detailed analysis of strategies for collaborative and multidisciplinary New Product Development (NPD) resulting in the design of three innovative medical products. The worldview of eight design participants are evaluated with interviews conducted throughout the design process. Each 'case' is analysed in turn before common characteristics and attributes are identified and associated with the stakeholder responsibilities, knowledge and motivations.

Bertola and Teixeira's (2003) analysis of 30 design case studies provides a detailed insight into the role design plays as a strategic competence for innovation in new product development. To do a comparative evaluation the study selected half the cases which were oriented towards international organisations operating in complex technologies, with the remainder being the local markets developing products utilising mature technologies. The selection of the cases and analysis of documentary evidence along with designers' own narratives allowed the capacity and influence of design on innovation to be established with common conditions.

Volker et al.'s (2008) 'Experimental/Intrinsic' case of the design of a public building highlights the inherent and irrational aspects of stakeholder's emotional response to perceived building and architectural quality against a seemingly rational and systematic decision-making framework.

Lastly, Yeomans' (1984) illustrative 'classic' case history describes the unique and innovative architectural features of the eighteenth-century roof at Lincoln Cathedral designed by James Essex. The case study is used to understand the structure of an unusually large roof span and the conceptual and structural characteristics associated with it to inform a 'natural history' of design.

Interestingly, only four articles in total were identified which sought to explain or demonstrate a design problem or product failure, such as Busby (2001) who investigated problems with the planning and design of a agricultural plant, and Petroski (1989) who sought to explore common design engineering failures in bridges.

However, aside from such examples, several common issues became evident which are problematic in the context of deriving theory from case studies. In summary these can be described as:

- variations in language and lack of specificity in terms used – the term 'case study' being used within the title, abstract or keywords with different use of terms throughout the article such as 'illustrative case/s', 'design case/s' and/or a 'case histories'

- process or method of case selection not described or justified

- methodology being ill-defined or in many instances non-existent, e.g. the phrase 'case study' is used as a descriptive label for a project located in a particular context but without reference to case study method

- lack of information to describe the processes used to derive theory, e.g. no audit trail to explain a conceptual 'leap'

- ontological issues associated with the method are not discussed or understood when diametrically opposing philosophical approaches are combined, e.g. inductive and deductive techniques and phenomenological and positivistic views of the world and how we experience it

- inability to explain the limitations of the 'case' for future scholarly research – for example the difficulty of transferability and/or generalisability of findings to other situations.

In short, problems with the context, research design and/or process of theory construction was evident in the vast majority of design case studies. In summary, this study has reinforced the need for design researchers to collate and interrogate contextual data surrounding the description of cases on two levels; within the external operating environment to describe the social, cultural and economic factors in play during the period of inquiry and the internal environment concerned with the resource, operational constraints and human perspectives involved. However, the illustration of the case-based designing approach also suggests that the traditional use of 'teaching cases' in design education may need to be reconsidered in terms of practice and interpretation.

Conclusions

There is growing interest in the quality and of design research in leading journals as the influence and contribution of design as a discipline internationally has grown in significance. Studies have sought to identify perceptions and determinants of quality across the design research community (see Friedman et al., 2008; Gerda et al., 2011), core themes in design research through conference contributions (Roworth-Stokes, 2011) and analysis of citations of papers in the journal *Design Studies* (Chai and Xiao, 2012).

This chapter has presented a systematic audit of design case studies in leading design research journals. The findings suggest there are significant variations in both subject coverage and theoretical issues covered. This raises issues for design case studies in an educational context. For example the number of case studies (roughly 8.5 per cent of the total articles reviewed) relative to discipline in areas such as fashion and graphic design (0 and 3 reported respectively out of 121) does not represent the undergraduate population (largest out of all creative arts subjects).

Furthermore, the research suggests that a common understanding of a teaching case in design pedagogy may not exist. Indeed the lack of a discipline language appears problematic and we discuss the issues raised in some detail, however we would also argue that the innovative interpretation and approaches to their use is also challenging and extending traditional pedagogic approaches within the field. Community of practice theory (Wenger, 1998;) suggests that disciplines will develop specific discourses, which are not easily understood externally. If students are to comprehend and benefit from this discourse they need a structured and responsive learning environment to master this shift in learner autonomy and subjectivity.

The illustrative case study has suggested that while there is limited literature surrounding the use of case studies in design education, there is unique and innovative pedagogy evident beyond the well-established case-based reasoning approaches evident in architectural design education. The approach to case-based designing serves to demonstrate cross-disciplinary practice and to question the role and purpose of design practice. Teaching materials sought to show examples of architects as product designers – either for integrated elements of their own buildings or as examples of independently commissioned product-scale works. In doing so, it raises the issue of micro versus macro perspectives on design. To provoke thought and debate to reinforce this, the group also viewed and discussed Cosmic Zoom, the animation short film by Eva Szasz (1968), that illustrates so succinctly the relationship between these subjects. This exemplifies the shifting roles and expectations of designers from generators of new products, services and experiences through to originators of innovative thinking processes as demonstrated within the illustrative case study.

As Osmond and Mackie (2012: 1404) explain, there are many difficulties when trying to transfer language and theories from other disciplines: they conclude that there are specific difficulties in the interpretation and motivation of creative arts subjects to adopt to tools and pedagogy from other disciplines:

> *Yet another reason could be, as with other studies carried out with design students using standard measurement tools, that most are not designed for students of creative subjects, not least because of the focus on text-based delivery methods ... From author experience, not only are design students more likely to respond to visual images, incidences of dyslexia occur more frequently in creative arts students ... creative students resist easy measurement due to the uncertainty that is a characteristic of creative thinking.*

To an extent the findings of this case are at odds with the discussion concerning case-based reasoning in architectural education. Here, the use of the 'case' becomes a means to interpret and understand how the designer becomes a facilitator of ideas and draws upon the design philosophy of others to mediate between design constraints and the conceptualisation of solutions to meet user requirements. It could be argued that it has been adapted and interpreted within industrial design as an extension of problem-based learning. It opens up new ways of exploring design practice and generating solutions.

In contrast to a developing literature on case-based reasoning (see for example Maher et al., 1995), this study found only a few cases used in architectural education that had been distilled into a repository of ready made 'recipes' to solve design problems. Clearly further research should be undertaken to categorise teaching case studies in design but these results imply that Breslin and Buchanan's (2008) call for case studies to bridge the gap between the development of theory and practice in design education may still be work in progress.

Acknowledgement

The authors are very grateful to Ben Kubler, Martyn Billings and Sean Mobsby for their support and involvement in the production of the illustrative case study.

References

Abaster. (2009) Originally cited from: http://www.abaster.com/2009/01/design-is-like-cooking-maybe.html.

Akin, Ö. (2002) Case-based instruction strategies in architecture, *Design Studies*, 23(4), 407–431.

Baynes, K. (1982) A case study in action research, *Design Studies*, 3(4), 213–219.

Bertola, P. and Teixeira, J.C. (2003) Design as a knowledge agent: how design as a knowledge process is embedded into organizations to foster innovation, *Design Studies*, 24, 181–194.

Bilda, Z., Gero, J.S. and Purcell, T. (2006) To sketch or not to sketch? That is the question, *Design Studies*, 27(5), 587–613.

Breslin, M. and Buchanan, R. (2008) On the case study method of research and teaching in design, *Design Issues*, 24(1), 36–40.

Burdick, A. and Willis, H. (2011) Digital learning, digital scholarship and design thinking, *Design Studies*, 32(6), 546–556.

Busby, J. (2001) Error and distributed cognition in design, *Design Studies*, 22(3), 233–254.

Bussracumpakorn, C. (2002) The study of the UK SMEs employing external organisations to support innovative products, in *Proceedings of the Common Ground Conference, Design Research Society*, D. Durling and J. Shackleton (eds) Staffordshire University Press, Staffordshire University. UK.

Chai, K. and Xiao, X. (2012) Understanding design research: a bibliometric analysis of design studies (1996–2010), *Design Studies*, 33(1), 24–43.

Chetty, S. (1996) The case study method for research in small- and medium-sized firms, *International Small Business Journal*, 15(4), 73–86.

Chiu, M.L. (2002) An organizational view of design communication in design collaboration, *Design Studies*, 23(2), 187–210.

Christiaans, H. and van Andel, J. (1993) The effects of examples on the use of knowledge in a student design activity: the case of the 'Flying Dutchman', *Design Studies*, 14(1), 58–74.

Cooper, R., Wootton, A., Hands, D., Economidou, M., Bruce, M., Daly, L. and Harun, R. (2002) Design behaviours: the innovation advantage – the multi-faceted role of design in innovation, in *Proceedings of the Common Ground Conference*, D. Durling and J. Shackleton (eds) Design Research Society, Staffordshire University Press, Staffordshire University. UK.

Cross, N. (2006) Forty years of design research, Presidential address, in *Proceedings of the DRS WonderGround Conference*, Lisbon, Portugal, 1 November 2006.

Dyer, G. and Wilkins, A. (1991) Better stories, not better constructs, to generate better theory: a rejoinder to Eisenhardt, *Academy of Management Review*, 16(3), 613–619.

Eisenhardt, K. (1989) Building theories from case study research, *Academy of Management Review*, 14, 532–550.

Friedman, K., Barron, D., Ferlazzo, S., Ivanka, T., Melles, G. and Yuille, J. (2008) *Design Research Journal Ranking Study, Preliminary Results*, correspondence with the author, 2008 July 26, Swinburne University and RMIT.

Gerda, G., Bont, C., Hekkert, P. and Friedman, K. (2011) Quality perceptions of design journals: the design scholars' perspective, *Design Studies*, 33(1), 4–23.

Goldschmidt, G. and Tatsa, D. (2005) How good are good ideas? Correlates of design creativity, *Design Studies*, 26(6), 593–611.

Horne-Martin, S., Jerrard, B., Newport, R. and Burns, K. (2002) Design, risk and new product development, in *Proceedings of the Common Ground Conference*, D. Durling and J. Shackleton (eds) Design Research Society, Staffordshire University Press, Staffordshire University. UK.

Jonson, B. (2005) Design ideation: the conceptual sketch in the digital age, *Design Studies*, 26(6), 613–624.

Langrish, J. (1993) Case studies as a biological research process, *Design Studies*, 14(4), 357–364.

Lawson, B. (1997) *How Designers Think: The Design Process Demystified*, 3rd edn. Woburn, MA: Architectural Press.

Lee, N. (2009) Project methods as the vehicle for learning in undergraduate design education: a typology, *Design Studies*, 30(5), 541–560.

Lehoux, P., Hivon, M., Williams-Jones, B. and Urbach, D. (2011) The worlds and modalities of engagement of design participants: a qualitative case study of three medical innovations, *Design Studies*, 2(4), 313–332.

Maher, M.L., Balachandran, M.B. and Zhang, D.M. (1995) *Case-Based Reasoning in Design*. New Jersey: Lawrence Erlbaum Associates, Inc.

Mau, B. (1998) *Incomplete Manifesto for Growth*, originally cited from http://www.brucemaudesign.com/4817/112450/work/incomplete-manifesto-for-growth.

Osmond, J. and Mackie, E. (2012) Designing for the 'Other', in *Proceedings of DRS 2012 Conference*, pp. 1396–1420. Bangkok Chulalongkorn, Thailand, 1–4 July.

Oxman, R.E. and Oxman, R.E. (1992) Refinement and adaptation in design cognition, *Design Studies*, 13(2), 117–134.

Petroski, H. (1989) Failure as a unifying theme in design, *Design Studies*, 10(4), 214–218.

Roworth-Stokes, S. (2006) Research design in design research: a practical framework to develop theory from case studies, in *Design Research Society Proceedings of Wonderground Conference*, November, IADE, Lisbon, Portugal, http://www.iade.pt/drs2006/wonderground/proceedings/fullpapers.html.

Roworth-Stokes, S. (2011) The design research society and emerging themes in design research, *Journal of Product Innovation Management*, 27, 485–496.

Sachs, A. (1999) 'Stuckness' in the design studio, *Design Studies*, 20(2), 195–209.

Spradley, J. and McCurdy, D. (1979) *The Ethnographic Interview*. New York: Holt, Reinhart and Winston.

Stake, R. (1994) Case studies, in N. Denzin and Y. Lincoln (eds), *Handbook of Qualitative Research*, pp. 236–247. Thousand Oaks, CA: Sage.

Strauss, A. and Corbin, J. (1990) *Basics of Qualitative Research: Grounded Theory Procedures and Techniques*. London: Sage.

Sugiyama, K. (2003) Results of the survey of education in design research, in *Proceedings of the 3rd Doctoral Education in Design*, Tsukuba, Japan, October 2003.

Szasz, E. (1968) *Cosmic Zoom*, directed and animated by Eva Szasz, 8 minutes, Canadian Film Board found at: http://www.nfb.ca/film/cosmic_zoom/related_films.

Theroux, P. (1999) *Sir Vidia's Shadow: A Friendship Across Five Continents*. Harmondsworth: Penguin Books.

Volker, L., Lauche, K., Heintz, J. and de Jonge, H. (2008) Deciding about design quality: design perception during a European tendering procedure, *Design Studies*, 29(4), 387–409.

Yeomans, D. (1984) Structural design in the eighteenth century: James Essex and the roof of Lincoln Cathedral Chapter House, *Design Studies*, 5(1), 41–48.

Yin, R. (1984) *Case Study Research: Design and Methods*. Beverly Hills, CA: Sage.

Yin, R. (1993) *Applications of Case Study Research*. Thousand Oaks, CA: Sage.

Chapter 11

Amplifying Learners' Voices through the Global Studio

AYSAR GHASSAN AND ERIK BOHEMIA

Introduction

Understanding the construction of autobiographical processes is an important aspect of gaining entry to the professional working environment. A central part of constructing such processes is learning how to tell appropriate stories about oneself to prospective employers. As such, this chapter argues design students must learn to tell their own stories.

Though it has tangible and important benefits for students in terms of aspects such as skills acquisition, this chapter argues that the commonly utilised master–apprentice model may not be optimally effective in aiding students to tell their own stories. Consequently, it may not be optimally attuned to enable future design graduates the necessary reflexivity to be able to negotiate the increasingly complex world of the contemporary knowledge economy.

The Global Studio aims to propagate a student-led pedagogic model in which tutors purposefully try to remain relatively distant in teaching and learning activities and students construct conversations and outcomes primarily via interaction with peers. Qualitative student feedback suggests that this model has enabled learners to tell their own stories. However, feedback also suggested that many learners are not comfortable with the notion that tutors remain relatively distant in the Global Studio system. Experiences from the Global Studio suggest that there is still much work to do in achieving an optimally balanced design education model.

The Master–Apprentice Model

An approach to design education which perceives the tutor as *master* and the student as his or her *apprentice* can be traced at least as far back as the staatliches Bauhaus. In the Bauhaus Manifesto and Program, the educational institution's founder Walter Gropius (1919: 1) decreed: 'there will be no teachers or pupils in the Bauhaus but masters, journeymen, and apprentices'.

Gropius' vision of the ideal educational scenario meant that the Bauhaus programme would consist of a tiered pedagogical model, the final stage being that for aspiring junior masters,

> *The training is divided into three courses of instruction:*

> *I. course for apprentices,*

> *II. course for journeymen,*

> *III. course for junior masters. Gropius (1919: 2)*

Though the rigidity of Gropius' tiered system seems inappropriate in the contemporary era, the top-down master–apprentice model remains extremely influential in design education (e.g. Tonkinwise, 2011). Reflecting on this established pedagogical system is not at all a straightforward matter because the discussion is multifaceted. This discussion will begin by examining some of the pressures which may be placed on contemporary design students.

Pressures on Design Students

When one takes into account tuition fees, the price of course materials, accommodation and the cost of living, the university experience for today's higher education students is likely to be an expensive one (see, e.g. Paton, 2013). Design students are of course not immune from experiencing financial difficulties associated with studying for a tertiary qualification. Their plight may be exacerbated by the existence of internship programmes which help to make up sandwich programmes of study. Dick Powell (Powell, 2013) – one half of the founding partnership behind the world-renowned product and automotive design consultancy Seymourpowell and chairman of the design charity D&AD – underscores the conundrum faced by design agencies when employing interns, 'We always try to pay a basic wage but it would clearly be better for [students] if we took on six that are unpaid than two that are paid, but we choose to pay them.'

For design students, financial strains may remain a feature of life beyond their time at university. Both authors have experience of teaching design at university level in a number of universities. Ghassan is privileged to have gained experience teaching both Transport Design and Industrial Design at higher education level. In his experience, students' passion for either (or both) of these areas is a major contributor in steering an overwhelming majority of learners towards embarking upon their degree programme. This fervour for the subject does not necessarily translate into well-paid graduate employment (Ball et al., 2010). Dick Powell (Dezeen, 2013) highlights the unenviable position occupied by contemporary design graduates:

> *Sadly, these days, it is harder than ever for graduates to find work; the jobs don't come to them – many don't realise that their graduation is the start of a lengthy, often soul-destroying process that is as much work as work itself.*

Powell's words may come as little consolation for students attempting to unleash their creative potential while studying for a design degree. In recent times, creativity has been a much-lauded trait for an individual to posses. It is viewed to be beneficial to society (Boden, 1999; Sosa and Gero, 2005) and the economy (Ball et al., 2010; Blair, 1998; Sands and Worthington, 2007) as well as being good for personal well-being (Fisher and Specht, 1999). However, researchers question whether enrolling on a *creative* subject at HE level aids graduates in finding meaningful employment. Design graduates belong to a group of individuals Comunian et al. (2010) term 'Bohemian Graduates': such scholars have been awarded tertiary qualifications in subjects such as creative arts, mass communications and music recording. The employment prospects for Bohemian graduates are less promising than those for their non-Bohemian counterparts (Comunian et al., 2010). Many find themselves in low-paid, mundane roles which are not suited to the qualification they have been awarded (Aston, 1999). Moreover, when Bohemian graduates are employed in creative occupations, their starting salary is lower than that of non-Bohemian graduates employed in creative roles by an average of almost £4000 (Comunian et al., 2010: 400). Research also suggests that the medium-term fate for graduates from creative subject areas presents a palpable level of concern. Hamish Coates and Daniel Edwards (2009: 15) claim that graduates in subject areas which include 'the creative arts … reported the lowest rates of full-time employment'.

Let us suppose that a particular undergraduate design student is aware of at least some of the arguments noted above. If so, it is understandable that they should worry about their employment and earning prospects. It is comprehendable then that a design student in today's climate should expect to be taught the necessary skills that will enable them to gain an advantage over their competitors in the design world. In the author's experience, design students expect (at least during the first half of their degree) that their tutor contact time at university should consist of experts in the field teaching them 2D visualisation skills (utilising both manual and digital tools) and 3D prototyping skills (physical and CAD modelling). In other words, in order to help them gain access to internships and future employment, students *expect the master–apprentice model to be propagated*. If the top-down educational system is what students demand, then what might be the grounds for critiquing it?

Criticism of the Master–Apprentice Model

Walter Gropius (1919: 1) underscores the notion that the master–apprentice system is a linear top-down model by pointing out that the master 'instruct[s]' the novice. Gropius (1919: 3) highlights the causality of the model by stating, 'the instruction of the individual is left to the discretion of each master'. If tutors are trained, passionate, experienced and proficient, what may be the harm in propagating this system in the contemporary era when students are eager to learn design skills? This chapter moves on to reflect on points of criticism aimed at this model of teaching.

First, rather than being student-led, the master–apprentice system is argued to be tutor-led. The layout of design studio teaching spaces is consistent with that of more generic teaching areas (JISC, 2006). According to JISC (2006: 10) the latter are traditionally 'tutor-focused, one-way facing and presentational, with seating arranged in either a U shape or in straight rows'. It is possible to suggest this tutor-focused arrangement may affect the continued development of *voice* in learners – for researchers have argued that while conducting studio teaching sessions, design educators talk more than their students are at the centre of learning activities (see Davies and Reid, 2000). Furthermore, the design teaching model is argued not to provide optimal conditions for creating mature relationships between students and tutors in the classroom (Baxter Magolda, 2009). For Jorge Frascara (2007) this approach curtails students' development, evidenced through their delivery of unimaginative forms. A perspectivist model is argued to be the best approach in art and design education (Danvers, 2003). In what appears the antithesis of this Frascara (2007: 64) states,

I have seen [design] instructors judge the quality of their students' work by saying: 'This one is too busy' or 'This is better, it is simpler.' They suggest that 'busy' is bad and 'simpler' is better in every situation'.

More ominously, the master–apprentice model may necessitate the need to question the authenticity of design decisions made by students. On this subject, Cameron Tonkinwise (2011: 452) argues 'design education is exemplarily Bourdieusian' in that tutors' values dictate outcomes delivered by students.

Second, researchers question whether the master–apprentice model prepares graduates for life as a professional designer. Problem-solving through linear, causal means remains the most widely utilised method of processing seen in design teaching (Findeli, 2001). However, rather than being easily definable, design problems are claimed to be complex (Lawson, 2006). Such problems are claimed to require a non-linear approach to tackle them (Buchanan, 1992). Alain Findeli (2001: 16) claims the tradition and dominant design teaching method has become obsolete: 'the canonical, linear, causal, and instrumental model is no longer adequate to describe the complexity of the design process.'

Designers are argued to be able to work at high strategic levels in their practice (e.g. Brown, 2009; Martin, 2009; Bevan et al., 2007). Brigitte Borja de Mozota (2011) disputes whether design education enables designers to operate optimally in tough professional climates. For Borja de Mozota the problem lies in the notion that even though designers 'have this potential to work at higher strategic levels ... they are not trained to do so'. This, she claims, 'is a challenge for design education' (Borja de Mozota 2011)

A third critique of the master–apprentice system is linked to the matter of graduate employment in the *wider* knowledge economy. Walter Powell and Kaisa Snellman (2004: 199) define the knowledge economy as 'production and services based on knowledge-intensive activities that contribute to an accelerated pace of technical and scientific advance, as well as rapid obsolescence'.

Trade in knowledge is increasingly important to the global economy (Lüthi et al., 2010). In developed capitalist economies, the production of knowledge is argued to be more important than any physical commodity (Drucker, 1993; Powell and Snellman, 2004). The knowledge economy is of relevance to design graduates as individuals in receipt of a university degree are 'purported to be the "knowledge workers" of the future and are expected to command high levels of general and specialist knowledge' (Brown et al., 2003: 109).

The ability to problem-solve is important for such individuals (Brown et al., 2002). Indeed, as opposed to workers who are not part of this club, graduates are 'given permission to think' in their professional life (Brown et al., 2003: 110). Of importance to this discussion is Powell and Snellman's (2004: 199) noted point that the knowledge economy implies an accelerated rate of advancement as well as prompt obsolescence. This implies that graduates who are successful in the knowledge economy will have to adapt and tackle problems in spaces that are forever shifting. The idea that graduates must be prepared for a rapidly changing professional climate is supported by the educational theorist Ronald Barnett (2000) who proposes that these individuals are entering a 'supercomplex' (p. 257) environment – a world that 'exhibits global features of challenge, uncertainty, turbulence, unquantifiable risk, contestability and unpredictability' (2000: 262).

A function of this unpredictability is the notion that contemporary graduates may find themselves in occupations for which they did not *directly* train at university. As a result, Bohemian graduates (a club which includes design graduates) can find themselves in what might be termed a broad church of creative industries, for example in media, advertising, and design (Comunian et al., 2010).

One might question how appropriately a master–apprentice-based teaching model in a subject such as industrial design might prepare graduates who (either through choice or necessity) broaden their horizons in the search for graduate employment? Consequently, Ghassan and Bohemia (2013) argue that the tutor-led master–apprentice model may not serve to optimally aid design students negotiate the complexity of the contemporary graduate working environment.

At this point, it is important to note that this chapter does not advocate the removal of the master–apprentice model in design education. As well as benefitting greatly from tutors recounting their experiences of industrial practice, learners have the opportunity to learn a myriad of practical skills intrinsic to the design process from them. Versed, experienced and passionate educators play a vital role in helping students understand, for example, the techniques of perspective drawing, the importance of achieving a sound quality of drawn line and the necessity of realising tension in curves and surfaces. Indeed, the acquisition of such skills seems difficult to imagine without the continued propagation of the master–apprentice system. In providing advice to graduates wishing to enter the profession, Dick Powell (Dezeen, 2013) emphasises the importance of skills acquisition in design practice,

Be really, really good at one thing. Be a star at one thing. Be an expert at one thing ... every business needs dedicated skills of different kinds – skills with tools, like Alias or Pro-Engineer, or skills at drawing, research, film editing, animating, budgeting, selling or whatever. Being a star at one thing can get you in, maybe not in the role you want, but at least you will be in and learning – after that, it's up to you.

Some of the skills Powell mentions (e.g. drawing; CAD) are those that are traditionally passed on via the master–apprentice system. Given the standing of Powell in the design community, perhaps there can be no more legitimate advocate of the already noted Bauhausian quest to create junior masters from novice individuals? However, objection to the master–apprentice system should be raised when design educators' involvement in teaching – or for that matter, assessment – dictates student outcomes. In such situations this chapter argues that via their creation of design proposals students are *telling educators' stories*. Instead, the authors argue for a balanced educational model where students are able to *tell their own stories*. This ability is crucial in differentiating would-be design graduates from their competitors in the field. Hinting at the importance of this trait, Dick Powell (Dezeen, 2013) argues design education 'makes [graduates] all more the same than different, so it's hard to stand out'.

This chapter moves on to discuss the relevance of storytelling to the fate of contemporary design graduates.

Storytelling and the Design Graduate

Humans traditionally use a host of tools as vehicles to aid narration. Poetry, for example, is argued to be one such method. For centuries it has been utilised to disseminate complex notions, convey knowledge and precipitate emotions (Grisham, 2006). The ability to tell stories is not just important for budding crafters of verse; it is significant in the quest for graduate employment.

Art and design graduates have stated that attaining a degree award alone was not enough in aiding them to gain entry into the professional workplace (Aston, 1999). The situation for these creative graduates is by no means unique. It seems to reflect a wider trend which positions a university degree as a 'given' perquisite of entering the professional employment market (Brown et al., 2002: 28). Consequentially, the quest for professional employability status in the knowledge economy relies on individuals entering into what Brown et al. (2002: 20) term a 'competition for credentials'. Over and above a degree classification, such credentials can include those gained through undertaking recognised training courses, the development of a pertinent network of professional contacts and the demonstration of desirable personal qualities. Gaining these credentials is not a guaranteed route to employment, but without them a would-be professional is not allowed to enter the race to gain professional employment (Brown et al., 2003). Ghassan and Bohemia (2011) argue that such credential-accruing actions belong to a set of activities Nod Miller and David Morgan (1993) term *autobiographical practices*. This term signifies the development of a personal identity which presents an individual in a manner deemed appropriate to a given circumstance. Pertinent to this chapter, autobiographical practices can be deployed when a protagonist is required to 'tell a story about themselves' (Miller and Morgan, 1993: 133).

As well as giving valuable insight into an individual's take on self-presentation, Miller and Morgan argue (1993: 133) autobiographical practices are important in informing observers 'about the circumstances under which such practices were deployed'. When it comes to searching for professional employment, practices adopted by the candidate may provide a great deal of information about the culture of the industry and of a particular company located within it. Brown et al. (2002: 28) argue that in order to increase their chances, graduates should be aware of their prospective 'cultural capital' in relation to the culture of the profession and/or organisation they wish to enter and should translate that in to 'personal capital'. As such, graduates should be versed in constructing appropriate autobiographical practices. Given the fact that employment rates and earnings for design graduates can be low, acquisition of such skills could be especially useful to design students (Ghassan and Bohemia, 2011).

In order to tell effective stories about oneself, a protagonist must acquire a sense of personal reflection. For Darren Cambridge (2008: 251) this process requires an individual to develop an 'individualized, reflexive identity'. Given the importance of the notion of self-presentation, it is perhaps surprising that little research exists on this area within the field of design. Tom Fisher (1997) stands out as an exception to this observation.

John Coldron and Robin Smith (1999: 714) have argued that an aspect of being a professional is the 'construct[ion] of a suitable identity'. Its formation is inextricably associated with one's community of practice (Lave and Wenger, 1991), for a protagonist constructs an identity via 'negotiating a position' (Blåka and Filstad, 2007: 62) with their community. Within a workplace, various factors facilitate this identity-building activity. These include the utilisation of dress

(Pratt and Rafaeli, 1997), the tone of voice one employs, mannerisms such as gestures and facial expressions as well as the posture one adopts (see Ashforth and Humphrey, 1993). Social scientists term the above *signs* (see Ashforth and Humphrey, 1993). Identity constructs can be utilised to differentiate a professional in one field from those not employed in that arena and can be deployed to relate one's status within one's community of practice (Coldron and Smith, 1999). Such signs are therefore assessed in relation to the cultural values and norms of a particular profession.

Designers are involved in the creation of suitable professional identities (Fisher, 1997). Within the world of design, certain signs may be used to narrate particular points about a protagonist to fellow practitioners. Such signs include accreditation from professional bodies such as the Chartered Society of Designers or the Industrial Design Society of America; an individual's affiliation with cultural design icons and/or sources of influence (Rodgers and Strickfaden, 2003); a person's style of sketching (Tovey et al., 2003); and their portfolio (e.g. Best, 2009; Coroflot, n.d.; Goldsworthy, 2009). These signs may be used to convey a designer's accreditation, what Ghassan and Bohemia (2011: 4) term a practitioner's 'design political persuasions' as well as their experience, skills, flair and contemporaneousness. Designers who successfully employ such signs can be thought of as individuals who make efficacious use of autobiographical practices (Ghassan and Bohemia, 2011).

Rather than being fixed, professional identities are in perpetual flux (Coldron and Smith, 1999). Thus while constructing appropriate identities a protagonist must be aware of contemporary prerogatives, trends and signs. As professional identities are in a constant state of emergence, the accumulation of credentials is not an activity practiced solely by those on the hunt for employment or promotion. It is instead vital in 'keeping fit' and in 'maintaining one's employability' (Brown et al., 2002: 24). Consequently, Ghassan and Bohemia (2011) argue an understanding of autobiographical practices may be beneficial to design graduates in their quest to gain employment in the knowledge economy. As has been noted, the ability to *tell stories about oneself* is integral to the construction of autobiographical practices (Miller and Morgan, 1993).

The authors argue that through possessing the capability of dictating students' design outcomes, the master–apprentice model does not best prepare individuals for *telling their own stories* to the design world. These stories have the potential of helping individual students to stand out amongst the design crowd. Furthermore, the notion that professional identities are forever in flux

(Coldron and Smith, 1999) serves to underscore the argument that design students' identities (and consequently the tools which help construct that – for example their drawing styles or portfolios) should be viewed as being in a constant state of emergence rather than being constructed through a top-down linear system. Through their experience as practitioners and educators, tutors can play a vital role in aiding students understand the necessity to grow their prospective cultural capital and how they may translate that into personal capital in the design world. As such, design educators play a central role in helping students to tell fabulous stories about themselves.

The remainder of this chapter discusses an educational initiative which differs from the tutor-led model commonly utilised in design pedagogy. Named the Global Studio, this model attempts to propagate a system in which lecturers purposefully try to remain relatively *distant* in teaching and learning activities and students construct conversations and outcomes primarily via interaction with peers. The authors move on to outline key features central to the Global Studio and to discuss successes and challenges related to running projects through it.

The Global Studio

The Global Studio utilises a blended learning approach – a combination of face-to-face teaching and online learning – to enable cross-institutional collaboration between universities located in international locations. Hundreds of students around the world take part in a Global Studio project. However, at a microscopic scale, assignments are conducted in the following manner: a small group of students in one institution utilises Web 2.0 technologies to collaborate with a small group of peers at a partner university.

Responding to the existence of globally networked organisations and the resultant shift in methods of working (e.g. Hoppe 2005; Horváth et al., 2003; Asokan and Payne 2008), the Global Studio enables small teams of design students to work with peers around the globe. Citing the relevance of the Global Studio, Neil Harrison and Nicola Peacock (2010: 878) claim it provides 'home students with [an opportunity to develop] a portfolio of globally relevant skills and knowledge without them leaving their home country'. Following in the tradition of the Design Studio, the Global Studio concentrates on project-based learning which is accomplished while and through doing (Schön 1985). This emphasis on project-based knowledge acquisition is argued to help embed established design practices into the repertoires of students (Bohemia and Harman, 2008).

In the Global Studio, face-to-face teaching takes place in two different ways. In the first phase, the layout of the design teaching room adheres to what JISC (2006) refer to as a traditional educational arrangement. Consequently, tutors stand at the front of the room and present information to home students who are seated in front of them. Tutors use this time to narrate schedules and logistical issues. They (and invited guests) also give talks related to the project theme as well as presentation techniques. Though students are given opportunities to ask questions, this period falls into the remit of being tutor-led. The second aspect of the face-to-face sessions involves tutors conducting group tutorials. In these, students have the opportunity to discuss project-related successes and challenges with tutors.

The online aspect of the Global Studio makes use of Web 2.0 technologies. A Global Studio WordPress site is created for each project. This features a homepage containing logistical information such as project themes, schedules and announcements. The homepage also relates information about the collaborating universities and participating tutors. Digital postcards from tutors and students are posted on this page. Within the Global Studio site, each pair of collaborating peers is provided with their own WordPress project sites through which they are able to communicate. Students are also free to choose to communicate via other Web 2.0 technologies such as Skype or Facebook. Tutors, other participating learners and industrial collaborators are encouraged to provide feedback to students via the WordPress project sites.

Tutors do not set briefs for Global Studio projects. Instead they develop project *themes*. The Global Studio attempts to deliver student-led projects which aim to prepare students for life as a professional designer. As in professional design practice, a student *client team* delivers a brief and a set of parameters for their collaborators, the student *design team*. As in professional practice, ultimately, the designers' task is to respond with an appropriate design intervention. In the Global Studio, client briefs and eventual design outcomes must exist within an overarching project theme provided by the academic project coordinators. This theme contains a set of deliverables as well as deadlines. Each team within the pairing performs both the client role and the designer role. Thus, Team A is the client for Team B. At the same time, Team B must write a brief and expects appropriate design interventions from Team A. It is important to note that when Team A acts as 'client', their brief contains instructions to design products or services that are relevant to an aspect of the culture in which they are 'home students'.

GLOBAL STUDIO PROJECT THEMES

Together, the authors have co-conceived and collaborated on two Global Studio projects.

Over 250 students collaborated on the first project. Named 'The Gift', this project was inspired by the eminent anthropologist Marcel Mauss' seminal book of the same name (Mauss, 1950). In an argument which has become a cornerstone of the social sciences, Mauss claims that *giving, receiving* and *reciprocation* are the central tenets of human interaction. The cultural theorist Stuart Hall (1997: 3) argues these interactions 'carry meaning[s] and value[s] for us, which need to be meaningfully interpreted by others'. The Gift project aimed to give students an appreciation of interpreting cultural practices intrinsic to the lives of their collaborator. Accordingly, it encouraged learners to explore the following aspects of communication and design:

- How do relationships form between people?

- How do bonds form between people of different cultures?

- Should cultural differences be bridged or should they be celebrated?

- What strategies might be employed in order to encourage relationships?

- What are the material effects of design? (Ghassan and Bohemia, 2011: 5)

The second project, entitled 'Festivals Fairytales and Myths', enabled over 200 students to experience working through the Global Studio. This project reflected the trend for authenticity in developed market economies (Arnould and Price, 1993). Facets of this trend include the development of the Slow Movement, resurgent culture and the growth in festivals and community events (Pietrykowski, 2004). The project presented an opportunity to highlight the significance of 'context' and 'meaning' to learners – for it is vital that practitioners are able to place designed artefacts and services in cultural and historical contexts. Peter Lloyd and Dirk Snelders (2003: 250) underscore this notion by stating that an object 'expresses or embodies ideas' in society. Paul du Gay et al. (1997) allude to the role of design practitioners in mediating cultural practices in arguing professionals 'play a pivotal role in articulating production with consumption by attempting to associate goods and services with particular cultural meanings' (p. 5) and are pivotal in presenting 'these values to prospective buyers'. Consequently, du Gay et al. (1997: 62) term designers as 'cultural intermediaries'.

As participating tutors, the contribution by the authors of this chapter to students' Wordpress sites consisted of posting encouraging remarks, positing questions about certain details learners had posted and reminding participating scholars of upcoming hard points in the schedule. Though informative and structured, to address the criticism of tutor-centred learning in design education, feedback provided by the above academics remained purposefully quite minimal in nature. Through creating an environment which centred on collaborative peer learning, both Ghassan and Bohemia wished to limit the overarching influence of tutors in this design teaching and learning environment. Through this, these academics aimed to *limit the likelihood of students telling tutors' stories*. Instead, through utilising Web 2.0 enabled cross-cultural peer collaboration, it was hoped that they would construct and narrate their own.

Students' Reflections and Discussion

Students were asked to provide feedback both at the mid-point and at the end of both of the Global Studio projects. The following qualitative reflection pertains to end-of-project feedback kindly provided by home students based at the UK institution. This data is pertinent as it allows insight into the whole of the Global Studio learning experience. Though this data has been conveyed in a previous publication (Ghassan and Bohemia, 2013), it is worth repeating students' observations here as they are pertinent to the discussion in this chapter.

As noted, the Global Studio aims to provide an opportunity for learners to appreciate that understanding cultures different from their own is important in contemporary design practice. Many students appeared to have gained insight into the relevance of this skill:

> *This festival is closely linked to Valentines Day, so it was important not just to skim over it and assume it as a Western celebration but look for the unique differences this day holds in China... To have a successful project I learned that it is highly important to spend time trying to empathise, understand and respect other people's cultures,*

and breaking through this barrier will ease communications and enhance productivity.

We had missed the point that in China cupcakes are not popular and don't hold the same meaning as in our Western culture.

Doing this project it has made me learn about other countries festivals and how they celebrate it.

It is important to understand cross-culture differences. And the differences should not be underestimated either.

Collaborating successfully with globally networked teams of professionals is an important feature of the contemporary knowledge economy. This involves a level of appreciation for how colleagues in other cultures see the world. Feedback suggested that for the majority of students, working with peers from cultures different to their own helped develop their intercultural sensitivity:

Learning to work with a design team from a different cultural background was challenging and interesting; it was all about learning about a new culture, having to both understand and respond to new, and different cultural cues.

Developing sensitivity for difference necessitates the critical evaluation of cultural stereotypes. Student feedback suggested that the Global Studio has helped learners critically evaluate cultural stereotypes:

Seeing/observing what the overseas team had found on our own culture (or my own) demonstrating what the cultural stereotypes were. What the overseas team found was not necessarily appropriate to our culture or reflected our culture, but based on these cultural stereotypes and clichés.

[I gained an appreciation of] the opinion of people so far away from the UK and Europe considering those places and how wrong are some stereotypes from both parties.

In the Global Studio, collaborating small teams must rely on a teaching approach that is not tutor-centred. Instead, collaborating students are co-dependent on one another's inputs. Individuals who felt they had benefitted from this learning experience noted they had learnt to rely on developing their own problem-solving strategies. Going beyond normal confines, self-evaluating design work, and feeling a greater level of control about their work's direction is as suggested by students' quotes below:

> *I had to go outside and experience [the] world. Get out of the shell that is the [class] room 103.*

> *We then had to go ahead and use our own judgement, as designers to decide as to what concept would work the best.*

> *[The project] created several challenges that needed to be addressed without input from lecturers. This definitely formed an environment that felt greatly independent of the university even though the project was undergone there.*

Bereft of the level of tutor-led teaching that is associated with the master–apprentice model, the feedback above suggests that many learners were able to construct and effectively narrate their own their own stories. However, paralleling this reflection upon the positive effect on learning, many students convey their struggle with making decisions without tutor-led involvement from educators. For example:

> *It would have been beneficial to the process if we could have had some input from the lecturers with regards to the actual designs too, perhaps resulting in some less dubious outcomes or smoother transitions between iterations.*

It is possible that feedback like this is related to the dominance of the master–apprentice model in design education. Below, feedback from a learner articulates how the normal way of working is administered by tutors – and how this affects the route a project may follow.

> *I have learnt an incredible amount from this project and they are things that I would never have experienced from the in-house projects at university, the projects we get from the university are regulated often by your tutors but it is so different when it is done by fellow students. Evidently our tutors are our clients and it's so easy to gain feedback and direction as they are there with you in your classroom however when working with international 'clients' it is clear to me how important communication is, how important leadership is and how*

> *communication your ideas in the right way can stop a lot of confusion*
> *and misunderstanding.*

A student who provided his feedback while conducting an internship with a manufacturer of expensive motor cars reflected the need for an approach to design education which is less *dictated* by tutors:

> *I feel I can understand this [Global Studio project] more so, as I've*
> *just spent my first week at the ... Design Studio, where its extremely*
> *fast-paced and not everything goes to plan when there are many things*
> *going at once.*

Conclusion

Professionals benefit from understanding the importance of autobiographical processes in acquiring and maintaining employability status in the contemporary working environment. Design students are it seems always a blink of an eye away from becoming fully fledged graduates. To improve their chances of gaining meaningful graduate employment, students must learn how to translate cultural capital into personal capital. To do this, design students need to learn how *to tell their own stories* to prospective employers.

Though it has tangible and important benefits for students in terms of aspects such as skills acquisition, this chapter has argued that the master–apprentice model may not be optimally effective in aiding students to tell their own stories. As the credentials for gaining and retaining employment are in a constant state of emergence, this model – which arguably can ask students to tell tutors' stories – may not be optimally attuned to enable future design graduates the necessary reflexivity to be able to negotiate the increasingly complex world of the contemporary knowledge economy.

This chapter argues for a balance between master–apprentice style teaching and a student-led approach to design education. Qualitative feedback has demonstrated that the student-led Global Studio has enabled learners to tell their own stories. Student feedback has also suggested that many scholars are not comfortable with the notion that the tutors remain relatively *distant* in the Global Studio system. Notes from the Global Studio suggest that there is still much work to do in achieving an optimally balanced design education model.

References

Arnould, E.J. and Price, L.L. (1993) River magic: extraordinary experience and the extended service encounter, *Journal of Consumer Research*, 20(1), 24–45.

Ashforth, B.E. and Humphrey, R.H. (1993) Emotional labor in service roles: the influence of identityemotional labor in service roles: the influence of identity, *Academy of Management Review*, 18(1), 88–115.

Asokan, A. and Payne, M.J. (2008) Local cultures and global corporations, *Design Management Journal*, 3(2), 9–20.

Aston, J. (1999) Ambitions and destinations: the careers and retrospective views of art and design graduates and postgraduates, *The International Journal of Art & Design Education*, 18(2), 231–240. doi: 10.1111/1468-5949.00179.

Ball, L., Pollard, E. and Stanley, N. (2010) *Creative Graduates Creative Futures*. Brighton: The Creative Graduates Creative Futures Higher Education Partnership and the Institute for Employment Studies.

Barnett, R. (2000) Supercomplexity and the curriculum, *Studies in Higher Education*, 25(3), 255–265. doi: 10.1080/713696156.

Baxter Magolda, Marcia B. (2009) Educating for self-authorship: learning partnerships to achieve complex outcomes, in C. Kreber (ed.), *The University and Its Disciplines: Teaching and Learning Within and Beyond Disciplinary Boundaries*, pp. 143–156. Abingdon: Routledge.

Best, J. (2009) Guest Post: A portfolio doesn't speak for itself, 21 August. Retrieved 20 January 2011, from http://www.coroflot.com/creativeseeds/2009/08/guest_post_a_portfolio_doesnt.asp.

Bevan, H., Robert, G., Bate, P., Maher, L. and Wells, J. (2007) Using a design approach to assist large-scale organizational change '10 high impact changes' to improve the National Health Service in England, *The Journal of Applied Behavioral Science*, 43(1), 135–152.

Blair, T. (1998). Foreword by the Prime Minister: in Our Competitive Future building the knowledge driven economy (p. 6).DTI. HMGO London.

Blåka, G. and Filstad, C. (2007) How does a newcomer construct identity? A socio-cultural approach to workplace learning, *International Journal of Lifelong Education*, 26(1), 59–73. doi: 10.1080/02601370601151406.

Boden, M.A. (1999). Computer models of creativity, in R. J. Sternberg (ed.), *Handbook of Creativity*, pp. 351–372. New York: Cambridge University Press.

Bohemia, E. and Harman, K. (2008) Globalization and product design education: the global studio, *Design Management Journal*, 3(2), 53–68. doi: 10.1111/j.1948-7177.2008.tb00014.x.

Borja de Mozota, B. (2011) *Design Economics-Microeconomics and Macroeconomics: Exploring the Value of Designers' Skills in Our 21st Century Economy* in Researching Design Education, pp. 17–40. CUMULUS/DRS Conference Proceedings, Paris, pubd CUMULUS, Aalto, Finland.

Brown, P., Hesketh, A. and Williams, S. (2002) *Employability in a Knowledge-driven Economy. Working Paper Series*, 38. Retrieved from Publications and Working Papers website: http://www.cardiff.ac.uk/socsi/research/publications/workingpapers/paper-26.html.

Brown, P., Hesketh, A. and Williams, S. (2003) Employability in a knowledge-driven economy, *Journal of Education and Work*, 16(2), 107–126. doi: 10.1080/1363908032000070648.

Brown, T. (2009) *Change by Design: How Design Thinking Transforms Organizations and Inspires Innovation*. New York: HarperCollins.

Buchanan, R. (1992) Wicked problems in design thinking, *Design Issues*, 8(2), 5–21.

Cambridge, D. (2008). Layering networked and symphonic selves: a critical role for e-portfolios in employability through integrative learning, *Campus-Wide Information Systems*, 25(4), 244–262. doi: 10.1108/10650740810900685.

Coates, H. and Edwards, D. (2009) The 2008 Graduate Pathways Survey: Graduates' education and employment outcomes five years after completion of a bachelor degree at an Australian university. *Higher Education Research*: Department of Education, Employment and Workplace Relations (DEEWR), Pub Australian Council for Educational Research, Melbourne, Australia.

Coldron, J. and Smith, R. (1999) Active location in teachers' construction of their professional identities, *Journal of Curriculum Studies*, 31(6), 711–726.

Comunian, R., Faggian, A. and Li, Q.C. (2010) Unrewarded careers in the creative class: the strange case of bohemian graduates, *Regional Science*, 89(2), 389–401. doi: 10.1111/j.1435-5957.2010.00281.x.

Coroflot. (n.d.). *Design Portfolio Tips*. Retrieved 20 January 2011, from http://www.coroflot.com/public/help_portfolio_tips.asp.

Danvers, J. (2003) Towards a radical pedagogy: provisional notes on learning and teaching in art and design, *International Journal of Art and Design Education*, 22(1), 47–57. doi: 10.1111/1468-5949.00338.

Davies, A. and Reid, A. (2000) Uncovering problematics in design education – learning and the design entity, in C. Swann and E. Young (eds), *International Conference on Design Education: Re-inventing Design Education in the University*, pp. 178–184. Perth, Australia: Curtin University of Technology, Curtin Print and Design.

Drucker, P.F. (1993) *Post-Capitalist Society*. New York: HarperBusiness.

du Gay, P., Hall, S., Janes, L., Mackay, H. and Negus, K. (1997) *Doing Cultural Studies: The Story of the Sony Walkman*. London: Sage Publications.

Findeli, A. (2001) Rethinking design education for the 21st century: theoretical, methodological, and ethical discussion, *Design Issues*, 17(1), 6–17.

Fisher, B.J. and Specht, D.K. (1999) Successful aging and creativity in later life, *Journal of Aging Studies*, 13(4), 457–472. doi: 10.1016/S0890-4065(99)00021-3.

Fisher, T. (1997) The designer's self-identity – myths of creativity and the management of teams, *Creativity and Innovation Management*, 6(1), 10–18.

Frascara, J. (2007) Hiding lack of knowledge: bad words in design education, *Design Issues*, 23(4), 62–68.

Ghassan, A. and Bohemia, E. (2013) From tutor-led to student-led design education: the global studio, in J. Beate Reitan, P. Lloyd, E. Bohemia, L. Merete Nielsen, I. Digranes and E. Lutnæs (eds), *Design Learning for Tomorrow – Design Education from Kindergarten to PhD: Proceedings of the DRS//Cumulus*

2nd International Conference for Design Education Researchers, pp. 542–536. Oslo, 14–17 May 2013. Oslo: AB Media.

Ghassan, A. and Bohemia, E. (2011) Notions of self: becoming a 'successful' design graduate, in N.F.M. Roozenburg, L.L. Chen and P.J. Stappers (eds), *Diversity and Unity: Proceedings of IASDR, the 4th World conference on Design Research*, 31, October–4 November, Delft, the Netherlands. Retrieved from: http://nrl.northumbria.ac.uk/12049/1/NotionsOfSelf723V20.pdf.

Goldsworthy, R. (2009) *Industrial Design Portfolio Advice: Back to Basics*. Retrieved 20 January 2011, from http://designdroplets.com/articles/portfolio-advice-back-basics/

Grisham, T. (2006) Metaphor, poetry, storytelling and cross-cultural leadership, *Management Decision*, 44(4), 486–503.

Gropius, W. (1919) *The Bauhaus Manifesto and Program*. Retrieved from http://www.thelearninglab.nl/resources/Bauhaus-manifesto.pdf.

Hall, S. ed. (1997) *Representation: Cultural Representations and Signifying Practices*. London: Sage.

Harrison, N. and Peacock, N. (2010) Cultural distance, mindfulness and passive xenophobia: using integrated threat theory to explore home higher education student's perspectives on internationalization at home', *British Journal of Educational Technology*, 36(6), 877–902. doi: 10.1080/01411920903191047.

Hoppe, R. (2005) The global toothbrush: international division of labor, *Spiegel: Special International Edition, The New World*, 130–135.

Horváth, I., Duhovnik, J. and Xirouchakis, P. (2003) Learning the methods and the skills of global product realization in an academic virtual enterprise, *European Journal of Engineering Education*, 28(1), 83–102. doi: 10.1080/0304379021000056839.

JISC (2006) *Designing Spaces for Effective Learning – a Guide to 21st Century Learning Space Design*. Retrieved from http://www.jisc.ac.uk/uploaded_documents/JISClearningspaces.pdf, accessed 2013.

Lawson, B. (2006) *How Designers Think: The Design Process Demystified*, 4th edn. Oxford: Architectural Press.

Lave, J. and Wenger, E. (1991) *Situated Learning: Legitimate Peripheral Participation*. New York: Cambridge University Press.

Lloyd, P. and Snelders, D. (2003) What was Philippe Starck thinking of?, *Design Studies*, 24(3), 237–247.

Lüthi, S., Thierstein, A. and Goebel, V. (2010) Intra-firm and extra-firm linkages in the knowledge economy: the case of the emerging mega-city region of Munich, *Global Networks*, 10(1), 114–137. doi: 10.1111/j.1471-0374.2010.00277.x.

Martin, R. (2009) *The Design of Business: Why Design Thinking Is the Next Competitive Advantage*. Boston, MA: Harvard Business Press.

Mauss, M. (1950, 1990) *The Gift*, translated by W. D. Halls. Abingdon, UK: Routledge. Original edition, Essai sur le don.

Miller, N. and Morgan, D. (1993) Called to account: the CV as an autobiographical practice, *Sociology*, 27(1), 133–143. doi: 10.1177/0038038593027001113.

Paton, G. (2013) Cost of a degree 'to rise to £26,000' after tuition fee hike. Retrieved from http://www.telegraph.co.uk/education/educationnews/10172015/Cost-of-a-degree-to-rise-to-26000-after-tuition-fee-hike.html, accessed 2013.

Pietrykowski, B. (2004) You are what you eat: the social economy of the slow food movement, *Review of Social Economy*, 62(3), 307–321. doi: 10.1080/0034676042000253927.

Powell D. (2013) Graduates should 'work for nothing' says D&AD chairman. Retrieved from http://www.dezeen.co013/07/17/graduates-should-work-for-nothing-says-d-and-ad-chairman/,in Dezeen Online Magazine

Powell, W. and Snellman, K. (2004) The knowledge economy, *Annual Review of Sociology*, 30, 199–220.

Pratt, M.G. and Rafaeli, A. (1997) Organizational dress as a symbol of multilayered social identities, *Academy of Management Journal*, 40(4), 862–898.

Rodgers, P.A. and Strickfaden, M. (2003) The culture of design: a critical analysis of contemporary designers' identities. Paper presented at the Design Wisdom: Fifth European Academy of Design Conference, Universitat de Barcelona, Spain. http://www.ub.edu/5ead/PDF/13/RodgersStrickfaden.pdf.

Sands, J. and Worthington, D. (2007) *High-level Skills for Higher Value*. London: Design Skills Advisory Panel and UK Design Industry Skills Development Plan.

Schön, D.A. (1985) *The Design Studio: An Exploration of its Traditions and Potentials*. London: RIBA Publications for RIBA Building Industry Trust.

Sosa, R. and Gero, J.S. (2005) A computational study of creativity in design: the role of society, *Artificial Intelligence for Engineering Design, Analysis, and Manufacturing*, 19, 229–244 doi: 10.1017/S089006040505016X.

Tonkinwise, C. (2011) A taste for practices: unrepressing style in design thinking, *Design Studies*, 32(6) 533–545. doi: 10.1016/j.destud.2011.07.001.

Tovey, M., Porter, S. and Newman, R. (2003) Sketching, concept development and automotive design, *Design Studies*, 24, 135–153. doi: 10.1016/S0142-694X(02)00035-2.

Chapter 12
Conclusions

MICHAEL TOVEY

This book is a product of design education research undertaken by members of the Design Research Society's Special Interest Group in Design Pedagogy. This group began in the UK but has established itself as an international group, chiefly through its biennial conference in design education research (DRS/CUMULUS in Paris in 2011, and Oslo in 2013, with the next planned for Chicago in 2015). Other research societies have similar strands of research in design education. The Design Society has an annual international conference in Engineering and Product Design Education, and the International Association of Societies of Design Research includes a strand dedicated to design pedagogy research.

It is quite natural that design academics should engage in such investigations, and that they should seek to extend our understanding and capability in this area. It is design academics who do almost all of the design research which leads to academic publications. Designers, for the most part, get on with designing, and leave design research to the academic community.

One of the key questions this book addresses is whether or not there are links between design research and design teaching. Clearly the conclusion is that there are such links, and maybe they could be closer. The strand running through the chapters is that design research does support design teaching, and they show a number of ways in which this is the case. This is a good reason for undertaking design research. If there is a close link with design teaching, particularly if design research supports effective design teaching, then that will gives design academics good reasons for doing such research.

Design research is not the same as research in some other disciplines. In a fundamental science such as physics if research stops then effectively the discipline comes to a halt. If there is no physics research then there is no physics.

Design is not like that. If academic design research were to stop then design would continue, more or less regardless. Designers would continue designing things, and probably the world would notice very little difference. It could be argued that design research is not central to design practice.

Much design practice includes a stage which is labelled as 'research'. It usually consists of the process of information-gathering to provide the starting point for designing, to inform the evaluative framework, and the context for the design. These are crucial parts of the process and essential to its success. However, this is not what we mean by design research.

Design research is an activity which is directed to exploring and understanding the nature of design, its processes and methods. It has loftier academic aspirations than the data-gathering part of the design process. It is usually undertaken by academics, and of course it is expected to conform to conventional standards of academic scholarship and rigour.

In universities and colleges there has been a long tradition of recruiting designers from design practice. However, the stronger tendency now is to regard the possession of conventional academic qualifications as a necessary prerequisite for holding a full-time academic position. Good practical experience is desirable but a Ph.D. is essential. In the context of the design discipline the clear implication is that to create a body of work for a Ph.D. in design you must undertake design research.

Design research is clearly necessary for the academic respectability of the discipline. However, it is not necessary for the actual activity of designing. As we have noted, if design researchers stopped doing design research, then design practice would continue, regardless. We have here the basis for a dangerous split in which we can see the practitioners regarding the design researchers, the design theorists if you like, as irrelevant, and unnecessary. And indeed in my experience this is exactly how some practicing designers do regard design researchers.

Design research does not just exist for its own sake. It is not there merely to comfort academics, as they shelter in an academic ghetto, shielded from the real world of design practice. It functions crucially to enable us to understand design better, and thus to enable design education to be improved. Where the research has a design practice focus then it also allows us to understand more deeply what is going on in the professional practice of design. That is one of the key themes of this book.

Design education research has taken a number of directions, focusing on the designer, the design context and the design interface, each of which provides a useful agenda for developing such research. Many see the end goal as that of achieving design programmes which are directed towards equipping graduates for entry to the community of professional practice. This in itself justifies the engagement of practitioners in the process. Various teaching strategies can accommodate these approaches. The studio, tutorial, library and crit are the traditional components, but using them effectively depends on the approach being informed by a deep understanding of the designerly way of knowing.

Many researchers reach the conclusion that the key to success lies in making arrangements which enable rather than inhibit the design student. On the face of it that is self-evident, and a statement of the obvious. But the natural emphasis on the peculiar nature of design thinking and the issues associated with wicked problems, and deploying tacit knowledge, can serve to overwhelm and demotivate. As circumstances conspire to erode motivation by swamping design tasks with information, this becomes all the more urgent. Design students need support for the agile navigation through the design process. Learning experiences should develop students' natural motivations and professionalise this motivation to create a resilient, informed and sustainable capacity. This is the essence of 'transformative learning', which is highlighted as the key to gaining entry to the community of practitioners. This can be approached through design learning based upon strengthening studio culture devised to provide safe spaces for creative and problem-centred learning and 'gateway' strategies of assessment. A key to sustaining motivation in this context is the toleration of design uncertainty as a threshold concept. It is argued that for student designers, liminal spaces can be unsafe places because they will not have the skills, experiences and confidence necessary to negotiate them successfully. Students need the time, space and structure to immerse themselves into a design brief, engaging in a reflective process to resolve the contradictions of a dual-processing cognitive model.

The community of practice notion is a major theme running through this book. The principle of engaging others to learn to become part of a community of professional practice underlies how most staff approach teaching and learning in design. Different design areas have their signature pedagogies and tutors support their students to become designers in ways which are specific to particular disciplines. Thus teachers may approach their teaching in qualitatively different ways, as it is perceived as contributing to engaging with the social practices which constitute the particular design practice. Participation in a community of practice is a key premise to understanding learning to practice, including learning the values and appropriating an identity related to that practice.

What engaging professional practitioners in the process does amongst other things is to allow informal access to a rich store of case-study material. Such an approach has greater effect if a greater body of case-study material can be accessed. From a wide range of examples it can be seen that it is possible to insert case studies in the curriculum. These can be used to identify and draw out design methods and approaches. It is important to include reflection by staff and students on the relevance and efficacy of this technique in informing a predominantly studio-based environment. Case studies have a potentially significant role in the development and acquisition of advanced, and professionally relevant, design skills and competences.

Of course in several contexts designers need to work in teams which may consist of the same types of designer, but are frequently cross-disciplinary. A most obvious pairing is that of industrial designers with design engineers. Each of these is an established community of professional practice with its own traditions, modes of communication and tacit knowledge. It is not unexpected in these circumstances for there to be barriers to communication during the development of a product. Design teachers need to investigate and understand why difficulties occur. Then they can develop a strategy to help resolve language issues and problems in understanding through a knowledge framework for design representations. Crucially if professional bodies can be involved then the approach can gain legitimacy.

Design teams can be international, and experience of working globally is an important component in the capability of a professional designer. Examples of industry-sponsored international collaborations between design students demonstrate how learners can address complex project situations and consequently prepare for contemporary working life. This is operationally different from 'tutor-led' design education as lecturers are more 'distant' in teaching and learning activities. Students construct conversations and outcomes primarily via interaction with peers. The intention is to prepare graduates for working in highly complex professional capacities synonymous with the contemporary era in which interactions with their peers are key to creating self-confident design graduates attuned to the contemporary community of design practice.

Thus we have an emerging range of well-researched proposals for templates for practice-based design education, aimed at producing graduates well suited to their various professional communities. Given the particular nature of the design disciplines there is a core need for the students to be enabled, and well motivated. In order to establish their identities as designers they will need to be able to tell their own stories. Such identities will relate to the particular signature characteristics and will depend on their having travelled through a transformative learning experience and overcoming the barriers which are particular to creative design practice. They will need to experience real-world design cases with the visual language which is characteristic of different types of designer. The developments reported on here demonstrate how this can be achieved.

Index